D0782256

SELECTED READINGS IN PHYSICS
General Editor: D. TER HAAR

COSMIC RAYS

COSMIC RAYS

BY

A. M. HILLAS

Reader in Cosmic Physics,
University of Leeds

PERGAMON PRESS

OXFORD · NEW YORK

TORONTO · SYDNEY · BRAUNSCHWEIG

Pergamon Press Ltd., Headington Hill Hall, Oxford
Pergamon Press Inc., Maxwell House, Fairview Park, Elmsford, New York 10523
Pergamon of Canada Ltd., 207 Queen's Quay West, Toronto 1
Pergamon Press (Aust.) Pty. Ltd., 19a Boundary Street, Rushcutters Bay, N.S.W. 2011, Australia
Vieweg & Sohn GmbH, Burgplatz 1, Braunschweig

First edition 1972
Library of Congress Catalog Card No. 70–181329

Printed in Hungary

08 016724 1

Contents

Preface

IT may appear surprising that the reprinted papers in Part 2 have not been taken from more recent work, but a short representative choice would have been difficult in a field where many basic questions remain unanswered. Instead I have looked further back, believing that readers will find interesting the ways in which physicists have tried to clarify these complex and only imperfectly understood phenomena. Part 1 fills in some of the more recent developments. Thus the original papers which are reprinted mark some of the turning points of one of the most fascinating stories in physics; but even so, in a book of this size it is not possible to cover the whole field of cosmic rays. I have therefore chosen to follow the thread which traces the subject from its beginnings towards the goal of finding where the radiation comes from. Although cosmic rays have been closely involved in the study of the elementary particles and the beginnings of high-energy physics, this aspect has been followed only to the point where the relationship between the primary and secondary radiations became understood in outline, so the work on most of the unstable particles, after the pion, is passed over. Also, on the subject of variations of cosmic rays, which, for example, is treated at length in *Cosmic Ray Physics* by Sandström, only one basic paper is included. Even so, because of the very extensive range of topics, it has been necessary to give a rather lengthy account in Part 1 to fill in the background to the papers reprinted in Part 2.

Although either Part 1 or Part 2 can be read separately, the papers of Part 2 are most logically read after the sections in which they are discussed in Part 1. Papers 1 to 10 treat the nature of the radiation, arranged chronologically; in Papers 11 to 16 the scene moves away from the Earth.

It would have been possible to make a very different choice of papers to reprint, as so many physicists have helped to advance the subject in many steps. One need only mention the very fine simple introduction *Cosmic Rays* by Rossi, in which he is able to draw very largely on his own work to illustrate the same story. However, the selection consists largely of papers which have had an important influence on the development of the field, and illustrate the methods by which the subject has been clarified. In several places, a short account has been given of the direction of later work following from the pioneering papers, leading on towards current work.

Some of the omitted topics may be found in the non-specialist accounts by Cranshaw, and Wolfendale, in books of the same title; for more advanced accounts of particular aspects, Ginzburg and Syrovatskii's book *The Origin of Cosmic Rays*, articles in the *Handbuch der Physik*, and in volumes of *Progress in Elementary Particle and Cosmic Ray Physics* (Wilson), and the proceedings of the IUPAP cosmic-ray conferences every two years may be consulted.

Although, in Part 1 only, MKS units are used in theoretical formulae, unless otherwise stated, numerical values are freely quoted in other units (ergs, gauss, cm) when these are customary and where the conversion involves simply powers of ten (though years and parsecs also appear).

Works are reprinted from the *Physical Review*, *Reviews of Modern Physics*, *Nature* (Macmillan Journals Ltd.), *Zeitschrift für Physik*, *Physikalische Zeitschrift*, *Il Nuovo Cimento* and the *Proceedings of the Koninklijke Akademie van Wetenschappen te Amsterdam*. I am very grateful to the publishers for permission to reprint the articles, and I am also very grateful to the living authors for their ready permission. I am responsible for shortcomings in the first two papers, translated from German originals, and I must thank Dr. E. W. Kellermann for helping me out on occasions.

I am also grateful to Professor P. L. Marsden for unpublished data on neutron monitors, used in two diagrams, and for much other useful information, and to him, Professor J. G. Wilson and Dr. C. J. Hatton for helpful discussions.

PART 1

1

Discovery

THE history of the investigation of cosmic rays is remarkable for the variety of the new phenomena revealed or illuminated, so one is prepared for some original turn of events whenever a new method or domain of observation is opened up. It is of interest to see this happening as we pass through the eras of Geiger–Müller counters, cloud chambers, emulsions, radio-telescopes, and so on, and as the site of the observations is extended. But first, we see what can be done with an electroscope.

1.1 The penetrating radiation

Around 1900 the phenomena of ionization and electrical conduction in gases were under investigation in several laboratories; the electron was a new discovery, the phenomena of radioactivity were becoming established, though without a fundamental understanding in terms of atomic nuclei, and the effectiveness of certain radiations in producing ionization in air was known.

Elster and Geitel in Berlin had pioneered the study of electrical conduction in the open air and had deduced that this resulted from the presence of positive and negative ions in the air. Attempts to interpret variations of the conductivity with geographical location, altitude, time and atmospheric conditions, and so to trace its source, were complicated by the fact that it depended not only on the concentration of ions but also on their variable mobility. This bedevilled many experiments for the next decade. However, Geitel and C. T. R. Wilson, also interested in atmospheric electricity, found that some ionization was present even in closed vessels containing air, even in the dark, Wilson concluding that 20 ion pairs must be formed per cm^3 per second. These conclusions were drawn from observations of the rate at which a charged gold leaf electroscope lost its charge even when protected against surface leakage. Wilson's account continues (Wilson, 1901):

"The experiments with this apparatus were carried out at Peebles. The mean rate of leak when the apparatus was in an ordinary room amounted to 6·6 divisions of the micrometer scale per hour. An experiment made in the Caledonian Railway tunnel near Peebles (at night after the traffic had ceased) gave a leakage of 7·0 divisions per hour, the fall of potential amounting to 14 scale divisions in the two hours for which the experiment lasted. The difference is well within the range of experimental errors. There is thus no evidence of any falling off of the rate of production of ions in the vessel, although there were many feet of solid rock overhead.

"It is unlikely, therefore, that the ionization is due to radiation which has traversed our atmosphere; it seems to be, as Geitel concludes, a property of air itself."

A worthy attempt—but it turned out (Elster and Geitel) that air in cellars and caves was especially conducting due to traces of radioactive emanation constantly seeping from the ground. It was clear that radioactive contamination played an important part in these observations. Perhaps all matter was to some extent radioactive.

Rutherford, then working in Montreal, himself observed the residual ionization in electroscopes, and found that some could be cut out by a lead shield (Rutherford and Cooke, 1903) indicating that part was due to γ-rays from the walls of his building, but inside a 5-ton mass of lead there remained 6 ions cm^{-3} s^{-1}. As late as the first edition of his book on radioactive substances (Rutherford, 1913) he concludes, however, that "the very weak activity actually observed is in all probability to be ascribed to the presence of traces of radioactive matter as impurities" —in the walls of the vessel. It may seem that Rutherford missed the mark, but the conclusion appears to be essentially correct, in so far as it refers to observations near sea level. Most of the effects observed were not due to an external radiation (which, it now appears, contributes one-quarter of Rutherford's residual ionization). It seems fortunate that most experiments around this time showed sufficient inaccuracy and unrepeatable results to stimulate further investigation over varying conditions and especially at higher altitudes. One observation in particular which spurred the search for an external radiation was a considerable diurnal variation in the ionization observed by several workers, pointing to the possibility of an extra-terrestrial source. However, the results were not reproducible: the variation was affected by atmospheric conditions and precipitation, and was much reduced as observing methods improved.

It should be remembered that the known ionizing agents arising from radioactive materials were α-particles, which leave a very heavy trail of ions along their tracks, and are stopped by $\sim$5 cm of air or a sheet of paper, β-particles, which expend their energy in ionization more gradually, and γ-rays, much the most penetrating, which would be cut down by a factor 10 in intensity by a few centimetres of lead, their ionization resulting only from secondary fast electrons ejected from materials they traverse.

To make more accurate measurements of the small amount of ionization Wulf developed a very sensitive electrometer in which the main part consisted of two quartz fibres fastened together at their ends, made slightly conducting, and carrying a small weight. When charged, the fibres bowed apart, and the very small electrical capacitance made the instrument sensitive to very small charges; and patient investigation showed that careful cleaning of zinc vessels gave less radioactive contamination and lower ionization rates.

This was the instrument used by most workers. It was possible to detect changes in ionization of a few ions $cm^{-3}\ s^{-1}$.

The ionization fell considerably over lake water or over a glacier (the experiments gave considerable scope for travel). As both should be relatively free from radioactive contamination, it became clear that much of the ionization arose from γ-rays emitted from radioactive trace elements in the earth, and quantitative estimates did not seem unreasonable. Although 3 ions $cm^{-3}\ s^{-1}$ remained over a glacier, and all estimates showed that the ionization to be expected from emanation escaping into the air should be at most one-tenth of the direct effect of γ-rays from the earth, it was hard to be certain that deposits did not form on the apparatus. However, Gockel did not find evidence of any decrease in the effect with altitude above a glacier. Experiments on towers showed a partial reduction as expected, but were inconclusive: in 1910 Wulf found that the ionization fell from 6 to 3·5 as he ascended the Eiffel Tower (330 m), whereas γ-rays should be halved in 80 m of air.

1.2 Manned balloon ascents

At this point, when much of the ionization had been traced to radioactive impurities and deposits, but the residual amount was uncertain, Hess in Vienna and Kolhörster in Berlin undertook the preparation of very remarkable ascents in open balloons to find how high the radiation penetrated.

Hess made two ascents to 1000 m in 1912, and after finding no drop in ionization realized that there must be a different cause. He then checked that at least up to 90 m from 1 g of radium C the γ-rays had an absorption coefficient of $4{\cdot}5\times10^{-5}\ \mathrm{cm}^{-1}$ in air. He also analysed carefully the sources of radioactive materials in the air, and found that they could account for at most one-twentieth of the ionization seen between 1 and 2 km up (Hess, 1913).

In September 1912 Hess reported at a meeting in Münster the results of seven balloon flights, now reaching over 5 km, in which he found that after an initial reduction as expected, the intensity of ionization became very much greater with altitude. He concluded that there was a very penetrating radiation coming through the atmosphere from above. This remarkable observation went almost unrecorded in the British scientific literature. Extracts from Hess's report are reprinted in Part 2 (Paper 1).

Kolhörster checked the effects which low temperature would have on the Wulf electrometers, and then in 1913 and 1914 made his dangerous ascents to 6 km and then 9 km, confirming and greatly extending the observations of an increase in radiation. Table 1.1 shows the average difference between the ionization observed at various heights (ions cm^{-3} s^{-1}) and that at sea level (Kolhörster, 1914).

TABLE 1.1

Altitude	1 km	2 km	3 km	4 km	5 km	6 km	7 km	8 km	9 km
Difference	−1·5	+1·2	+4·2	+8·8	+16·9	+28·7	+44·2	+61·3	+80·4

There was a radiation from space which had an attenuation coefficient of only 1×10^{-5} cm^{-1}, much more penetrating than known γ-rays.

(Extrapolating Kolhörster's measurements back to ground level, they suggest an ionization $2\frac{1}{2}$ ions cm^{-3} s^{-1} is attributable to the penetrating radiation there, and a little more to radiation from the ground.)

Now that there was clear evidence of an external radiation, the next problem was to locate its source. (It still is.) Hess referred (Paper 1) to the absence of any noticeable drop during a partial solar eclipse which, taken with the absence of a large day–night variation, pointed away from the Sun as a source. Many physicists kept observations of the ionization (screened from local γ-rays) for long periods of time in the hope of detecting variations with sidereal time, as the observer faced different parts of the sky. Although variations due to atmospheric changes came to light, and many observers claimed even very large daily changes, no real effects came out of this work, and it will be passed over.

1.3 Millikan's experiments

Not all physicists were convinced by the evidence of a penetrating radiation, however. In America, Millikan was apparently unimpressed in face of the variations amongst the results reported by different workers; and when he moved to the California Institute of Technology after the First World War he set out very energetically to obtain much more precise measurements of the ionization and its variation with altitude.

Millikan's advances in technique were very important, though at first his interpretation of the results confirmed his doubts. In 1922 Millikan and Bowen adapted meteorological sounding balloons to carry light-weight electroscopes which reached 15 km altitude, recording results automatically by casting a shadow of the fibres onto a moving photographic film. Two out of four balloons were recovered after their flights, and the results indicated much less ionization above 5 km than expected on an extrapolation of Kolhörster's observations, though a large effect of temperature prevented detailed analysis. In the following year an expedition to Pike's Peak (Millikan and Otis, 1926) confirmed his doubts when at an altitude of 4·3 km on the mountain he found the rate of ionization increased by 11·6 units compared with Pasadena, but if shielded by 48 mm of lead, the apparatus saw an increase of only 2·2 units. Millikan concluded that there was no cosmic radiation of the characteristics previously described, the increase at altitude representing a local γ-radiation. (It appears that the increase of 11·6 is what would be read from Kolhörster's balloon observations, contrary to Millikan's belief: the unexpected feature is really that lead is such an effective screen for radiation which is very penetrating in air.) From this it will be gathered that wrong conclusions can readily be drawn from cosmic-ray experiments unchecked by different techniques or different physicists.

In 1925 Millikan and Cameron began a series of measurements of the ionization produced in electrometers lowered into snow-fed lakes in California, which were thought to be free from radioactive contamination. The purpose was to clear up inconsistencies in measurements of the absorption coefficient of the radiation by making very precise measurements with a low background ionization. These results removed all doubts: in Muir Lake at an altitude of 3600 m the ionization dropped from 13·3 ions at the surface to 3·6 at a depth of 15 m; and in Arrowhead Lake at 2040 m the same readings were obtained at depths 6 ft less than those in Muir. As the mass of air between these altitudes corresponds to 1·85 m of water, the results strongly indicate a radiation

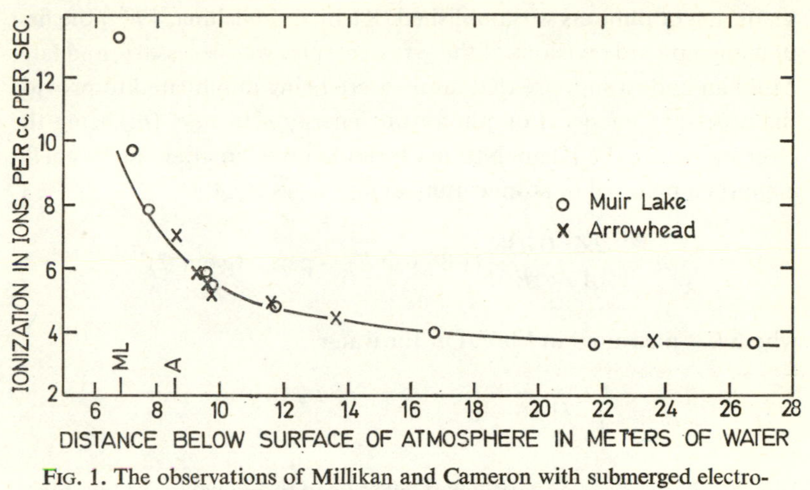

FIG. 1. The observations of Millikan and Cameron with submerged electrometers. ML, A, mark the lake surfaces.

travelling downwards through the air with no local generation. The results are shown in Figure. 1.

The absorption coefficient found for this radiation in the water was even smaller than had been suspected from the measurements in the open atmosphere, and the absorption was not exponential, but became steadily slower at greater depths. Millikan made repeated attempts over the next decade to analyse such absorption into a superposition of exponentials corresponding to γ-rays of different energies. The hardest rays on the graph shown, with an absorption coefficient of 0·18 m^{-1} water, were then taken to be γ-rays of $\sim$30 MeV, and later under 70 m water the coefficient was down to 0·04 m^{-1} water.

Millikan introduced the name "cosmic rays" at this point, and he already saw that they carried to the Earth an energy at least a few per cent of that arriving from the stars. Speculating on their possible origin, he suggested the synthesis of elements from hydrogen in space as the only adequate energy source, as by a remarkable coincidence the apparent energies deduced coincided with the nuclear binding energies of the common elements He (29 MeV), O and Si ($\sim$240 MeV).

When a more accurate quantum mechanical treatment of Compton scattering of photons was published (Klein and Nishina, 1929) the first of many upward revisions of the γ-ray energies was necessary, and later Millikan had to suppose that atoms were being annihilated to provide the necessary energy. For photons of energy $E \gg m_e c^2$ (m_e being the electron mass), the Klein–Nishina formula gives an attenuation coefficient (in a material of atomic number and mass Z, A)

$$\mu = \frac{2Z}{A}\,\frac{0{\cdot}038}{E}\,(1{\cdot}86+\ln E) \quad \text{per} \quad (\text{g cm}^{-2})^{\dagger}$$

where E is measured in MeV. Or, for water,

$$\mu = \frac{4{\cdot}3}{E}(1{\cdot}86+\ln E)\ \text{m}^{-1}.$$

But this theory of attenuation had never been tested beyond a few MeV, and in fact is incomplete: for no γ-ray energy is the total attenuation in water less than $1{\cdot}7\ \text{m}^{-1}$, a minimum reached at 40 MeV. In a lead absorber pair production has a dominant effect at even lower energies, but the anomalous absorption of the softer component in lead (i.e. not simply dependent on the density of atomic electrons in the material) was not appreciated and Millikan regarded observations of the less penetrating rays with suspicion. For this reason he did not discover any geographical dependence, when he made observations in a Bolivian lake. Perhaps it is fortunate that he could not detect a discrepancy in intensities at the two lake altitudes, an effect later exploited in connection with the instability of the particles, or the force of his observations might have been obscured.

Another fact which became better established during this period was that the radiation came from no preferred directions in space.

From Millikan's more precise measurements, it was deduced that at sea level the part of the ionization due to cosmic rays amounted to only 1·4 ion pairs $\text{cm}^{-3}\ \text{s}^{-1}$, in air at atmospheric pressure. It seems, therefore, that it did first make an undisputed appearance in Hess's observations.

† Mass per unit area is a useful way to measure thickness of absorbers.

II

The Nature of the Radiation

PAPER 2 takes up the question of the nature of the radiation which had been discovered.

The outstanding property of the radiation was its penetrating power: it was still detectable after passing through 1000 g cm^{-2} of our atmosphere, equivalent in mass to 76 cm of mercury. As we have seen, for the softer (less penetrating) part of the radiation, this did not seem to call for too large an extrapolation of the penetrating properties of known γ-rays, in contrast to electrons, or β-rays, which normally had ranges of < 1 g cm^{-2}. A plausible picture was thus of a γ-ray entering and penetrating a large part of the atmosphere before giving up most of its energy by elastic collision to an atomic electron. These recoiling electrons, of relatively short range, were seen as the agents causing ionization in detectors, through which the γ-rays were indirectly observed.

The likely existence of different "primary" and "secondary" rays was thus recognized at the start; and their inter-relationships soon set an unexpectedly challenging and tortuous problem. It is convenient now to consider firstly investigations into the nature of the radiation found near sea level, and then deductions which could be made about the "primary" radiation entering the atmosphere.

2.1 The observation of individual cosmic rays

In 1929 Skobelzyn, in Leningrad, had set up a small cloud chamber between the poles of a magnet to measure the energies of β-particles of radioactive origin from the curvature of their tracks. In some of the photographs he saw extraneous tracks which were hardly deflected, looking like electrons of energies greater than 15 MeV: these were the first pictures of tracks of cosmic rays, as he rightly recognized, noting that their rate of occurrence would account roughly for the amount of ionization normally attributed to the penetrating radiation.

In particular, he interpreted the particles as the Compton recoil electrons secondary to the "Hess ultra-γ-radiation", as did Auger with whom Skobelzyn then collaborated in further experiments.

However, a more convenient approach opened up in the same year following the invention of the Geiger–Müller counting tube, which permitted the counting of individual ionizing particles passing through the detector. Many of the impulses would be due to radioactive contamination within the counter, but Bothe and Kolhörster (working in the same institute as Geiger and Müller) observed that when two such counters were placed one above the other simultaneous discharges in the two occurred very frequently, indicating that a charged particle of sufficient penetrating power had passed through the two. This use of detectors discharging "in coincidence" has proved one of the most fruitful techniques in cosmic-ray and elementary particle physics; but whereas nowadays electronic circuits select coincident signals, Bothe and Kolhörster had to inspect film records.

Their classic experiment, which established the gross features of the cosmic rays near sea level, is described in Paper 2.

This was the first investigation in which the essentially corpuscular nature of the rays emerged, from the unexpected frequency with which simultaneous discharges were seen in the two counters. (In fact the proportion of coincident counts was probably underestimated somewhat, as their counters seem not to have been perfectly efficient, judging by the flux they found.) A Compton recoil electron might pass through two unshielded counters, but Bothe and Kolhörster found that coincident discharges were still quite common when 4·1 cm of gold was placed between the counters: in fact they showed that the charged particles seemed to be absorbed in matter at the same rate as the previously observed radiation, strongly suggesting that no more were being generated (by the hypothetical γ-rays), so that the cosmic radiation was composed mainly of these particles. The authors realized that energies of 10^9 to 10^{10} eV would be needed to give charged particles such long ranges in matter, and hence they might have unfamiliar properties. Their duly cautious argument may be read in the reprinted paper, although it is not reproduced in full.

One sees from Figure 4 of their paper that about three-quarters of the particles at sea level seem to belong to a distinctly "harder" component. Rossi later extended the observations to show that most of the particles which were not stopped in a few centimetres of lead could penetrate a metre of lead. These were certainly not secondaries knocked on by photons absorbed in the air. Further experiments were facilitated by Rossi's (1930) electronic coincidence detecting circuit, the modern transistor version of which is the basis of logic circuits (as the NOR circuit).

2.2 The latitude effect

Having found charged particles reaching the ground from a great altitude, Bothe and Kolhörster pointed to the likely effect of the geomagnetic field on particles of the energy in question if they approached from infinity. With the aurora in mind, they followed up their suggestion with an expedition to Spitzbergen, but found no greater intensity of cosmic rays there than in Potsdam. Millikan in 1930 made measurements at Churchill, only 730 miles from the north magnetic pole, but despite the appearance of auroral particles overhead his measurements agreed to 1% with those in Pasadena (magnetic latitude 41°). (An earlier measurement in Lake Miguilla, Bolivia, had also given an intensity which seemed normal, though the softer component of radiation, which Millikan ignored, appears to have been weaker in Bolivia.)

Nevertheless, beginning in 1927, Clay had come upon evidence in favour of a variation with latitude. This came about when he travelled home to Holland from his post in Java, where he had been studying altitude and time-variations of cosmic rays with ionization chambers and encountered the usual difficulty in determining the residual inherent discharge rates of the instruments, so he took a detector back to

Europe "for the purpose of making a determination in a rock-salt mine, which might be expected to be quite free from radiation.

"On the way a series of measurements was made on the boat-deck of the steamer Slamat, about 14 metres above the sea level, from the Red Sea to and partly including the Mediterranean. It is a remarkable fact, that an increasing value of the radiation was stated, which may be partially the result of increased gamma radiation of emanation in the atmosphere.

"It might have been that the residual radiation in the apparatus increased. As will presently appear, this is, however, unlikely."

The salt mine was a failure, but the voyages were repeated, and a decrease of 14% on going from Holland to Java was confirmed. At first Clay took this to indicate that the rays were of local, terrestrial, origin.

However, by 1932 (selected as an international "magnetic year") several groups of people compared measurements made in different parts of the world: most notably A. H. Compton had organized an expedition around the Pacific (and up mountains where available), and found near the equator a clear decrease of intensity, which affected the less penetrating rays more, and also noted from accurate measurements that the intensity levels followed lines of magnetic latitude more closely than geographic latitude. Clay realized that his observations varied steadily with geomagnetic field strength. It was then clear that a large proportion of the primary cosmic rays must be charged particles, some of which were deflected away by the stronger horizontal component of the magnetic field near the equator.

In Paper 3, Clay analyses the evidence regarding the nature of the cosmic rays, and concludes that the primary radiation consists of charged particles having energies ranging from below 4×10^{9} eV to well above 3×10^{10} eV. He also explained the failure to detect an increase in radiation at high latitudes, where particles of lower energy should be able to penetrate the field, by showing that they could not penetrate the atmosphere. This is true, but not the whole truth.

Lemaître and Vallarta (1933) published an approximate mathematical analysis of the influence of the Earth's field on charged particles, which served as a basis for a more quantitative understanding of the effect, although at first they did not take account of the effect of the atmosphere in shielding from inclined rays.

2.3 The effect of the Earth's magnetic field on charged particles approaching the Earth

Lemaître and Vallarta, and also Clay and others, made use of the earlier calculations of Størmer, who had developed a theory of the origin of the aurora borealis many years previously, and had shown that there were certain regions around the Earth inaccessible to charged particles of a given energy approaching from a distance. Størmer's theory was then applied to cosmic rays, on the assumption that they were charged particles approaching the Earth from all directions.

Charged particles of sufficiently high energy would only be deflected slightly by the geomagnetic field, but particles of lower energy could have very complex trajectories, and although no general analytical description has been found, Størmer showed that they must lie within certain limits.

The Earth's magnetic field is approximately that of a dipole of moment M ($= 8{\cdot}1 \times 10^{22}$ A m^2), for which the components of magnetic induction at a distance d from the dipole centre and at a latitude λ are, along radial and transverse directions respectively,†

$$B_r = \frac{-2\mu_0 M \sin\lambda}{4\pi d^3} \quad \text{and} \quad B_\lambda = \frac{-\mu_0 M \cos\lambda}{4\pi d^3},$$

the latter in the direction of increasing λ. A force $q\boldsymbol{v}\times\boldsymbol{B}$ acts on a charged particle of velocity $\boldsymbol{v}$ and charge q: this does not alter the magnitude v of the velocity.

The simplest (but unstable) orbit for a particle of momentum p is a circle of radius b in the geomagnetic equatorial plane, where

$$b = \sqrt{\left(\frac{\mu_0 M q}{4\pi p}\right)},$$

† MKS units are used here: $\mu_0 = 4\pi \times 10^{-7}$.

and there are other looped orbits which never escape from the vicinity of the Earth: they never extend beyond the distance b. This critical distance b is taken as the unit of distance by Størmer, in order to simplify the equations of motion. r is the distance of the particle P from the dipole centre O (see Figure 2), measured in these units (which depend on the momentum of the particle, which does not change).

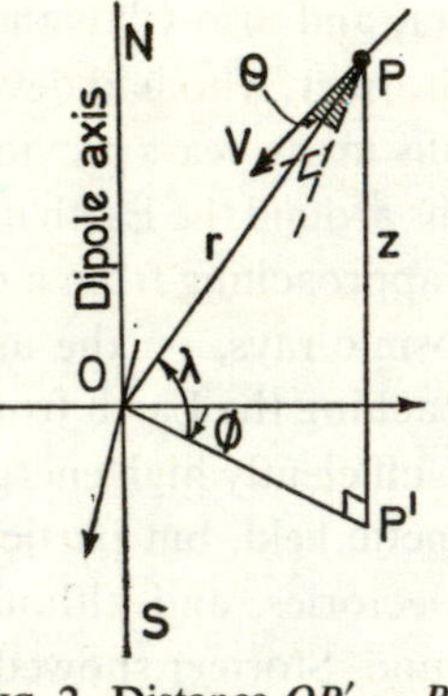

FIG. 2. Distance $OP' = R$.

The motion of the particle P is then described by giving the coordinates R, z (or r, λ) in the meridian plane OPP', and the azimuthal angle φ of this plane, whereupon the equations of motion (using the Størmer units for distance and velocity) may be written as follows.

(a) For motion in the (moving) meridian plane:

$$\ddot{R} = \frac{v^2}{2}\frac{\partial Q}{\partial R}; \qquad \ddot{z} = \frac{v^2}{2}\frac{\partial Q}{\partial z}; \qquad \dot{R}^2+\dot{z}^2 = v^2 Q;$$

where $Q = 1-\left(\dfrac{2\gamma}{R}+\dfrac{R}{r^3}\right)^2$, γ being any real constant.

(b) For the motion of the meridian plane:

$$\dot{\varphi} = v\left(\frac{2\gamma}{R^2}+\frac{1}{r^3}\right).$$

Størmer also showed that the angle ϑ between the trajectory at P and the meridian plane (positive if moving to the west) obeys the equation

$$\sin \vartheta = -\frac{2\gamma}{r \cos \lambda} - \frac{\cos \lambda}{r^2}. \tag{2.1}$$

Although no general analytical solution of these equations has been found, the fact that the quantity

$$2\gamma = -r \cos \lambda \sin \vartheta - \frac{\cos^2 \lambda}{r} \tag{2.2}$$

is a constant of the motion has been used to prove that trajectories can exist only within certain limits. (For orbits which do come from a great distance, 2γ has a simple meaning as an impact parameter: if the distant trajectory, inclined at an angle λ to the equatorial plane, were extended as a straight line it would miss the dipole axis by a distance $2\gamma/\cos \lambda$.)

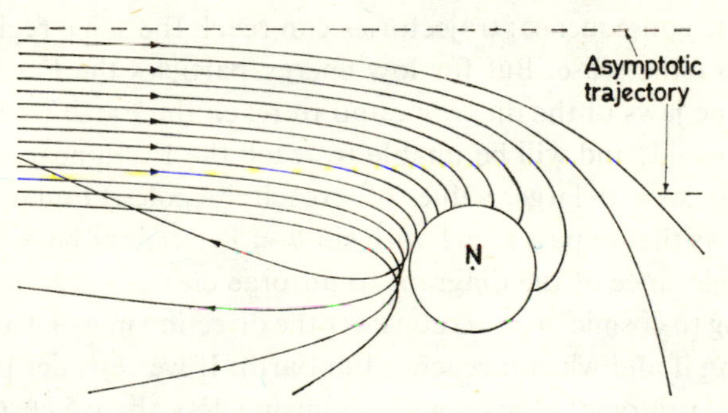

FIG. 3. Trajectories of 20 GV positively charged particles in the equatorial plane.

Figure 3 shows orbits for a series of impact parameters for the simple case where the trajectories lie in the equatorial plane ($\lambda = 0$). When γ becomes < -1, particles suddenly cease to be pulled in within the critical distance of 1 Størmer unit.

For the more general case of orbits not lying in the equatorial plane, the motion in R and z may be depicted, as in Figure 4b. The motion must lie in an unshaded area of the diagram, since Q must be positive (the

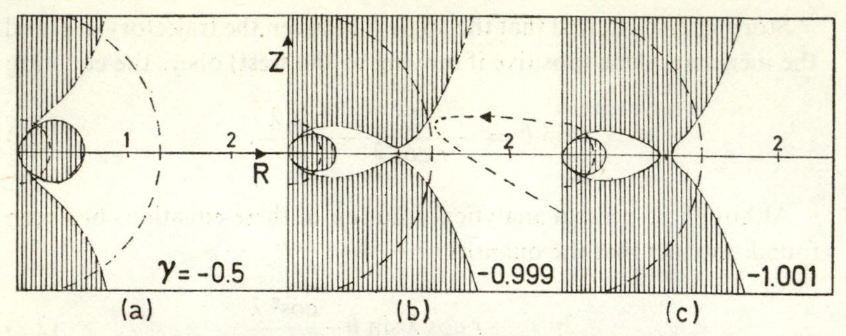

FIG. 4. Forbidden regions of the $R-z$ plane for trajectories with γ as shown. The dashed circles show the Earth for rigidities 100 GV (large circle) and 5 GV.

same limitations arising by requiring $|\sin\vartheta| \leqslant 1$), and the path in fact is the same as for a particle rolling in a potential trough, such that its kinetic energy is proportional to Q. For impact parameters near -1 a dramatic change occurs again. For high energy particles the Størmer unit is small, so the Earth appears large on the diagram and the change is of little consequence: trajectories can reach the same regions of the Earth in either case. But for low energy particles the Earth lies well within the jaws of the diagram, and to reach the Earth particles must have $\gamma > -1$, and will be unable to reach the Earth near its equator. (Trajectories with larger values of γ reach the poles.) Equation (2.2) in fact shows that to pass $r = 1$ with $\sin\vartheta \leqslant 1$, γ cannot be < -1.

The relevance of the diagram to aurorae clearly suggests itself, but returning to cosmic rays, we consider the direction in which the particle is moving if and when it reaches the Earth. If we consider particles of unit (i.e. protonic) charge with momenta less than 59·6 GeV/c the radius of the Earth is less than 1 Størmer unit, so that the impact parameter $\gamma \geqslant -1$ to reach the Earth, and so from (2.1),

$$\sin\vartheta \leqslant \frac{2}{a\cos\lambda} - \frac{\cos\lambda}{a^2},$$

a being the Earth's radius in these units. This means that all trajectories reaching an observer at geomagnetic latitude λ must lie within a cone of opening angle $(90°+\vartheta)$ about the horizontal line pointing west (for positively charged particles) or east (for negatively charged particles). This is Størmer's allowed cone. The other directions constitute a forbidden cone, illustrated (for positive particles) in Figure 5. Most important will be the particles arriving vertically, as they have less

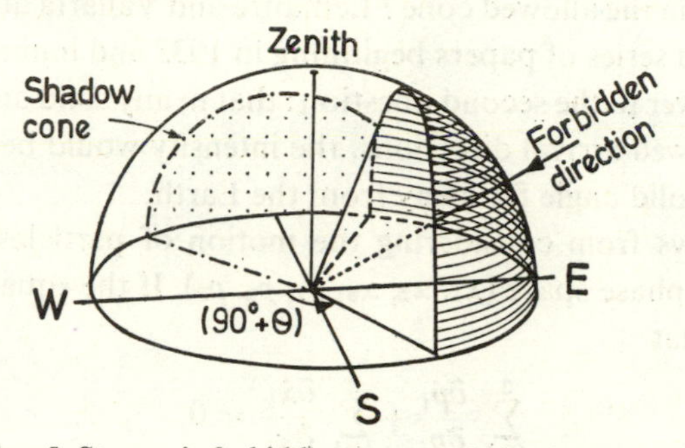

FIG. 5. Størmer's forbidden cone, and the shadow cone.

atmosphere to penetrate. The allowed cone includes the vertical direction if ϑ reaches 0°,

i.e. $$0 \leqslant \frac{2}{a\cos\lambda} - \frac{\cos\lambda}{a^2},$$

giving $$a \geqslant \tfrac{1}{2}\cos^2\lambda.$$

Substituting for a, a necessary requirement for a particle with momentum p and charge ze to arrive vertically is that

$$cp \geqslant z\cos^4\lambda \times 14{\cdot}9 \text{ GeV}.$$

Thus protons or electrons of momenta greater than $14{\cdot}9\cos^4\lambda$ GeV/c may enter vertically: at the equator the threshold would be 14·9 GeV/c, at latitude 40° it would be 5·1 GeV/c, at 60°, 0·93 GeV/c. (For protons, corresponding kinetic energies are 14·0, 4·3 and 0·48 GeV.) For particles of greater charge, threshold momenta would be higher: but the ratio (cp/charge), known as magnetic rigidity, would be the same—14·9, 5·1 and 0·93 GV.

This is as far as Størmer's theory went, and Clay used it to deduce the energies of those incident particles which were no longer detected as one went towards the magnetic equator.

However, two questions remained before one could have a quantitative theory. Størmer's theory showed really which directions of arrival of particles were forbidden: were all others allowed? And then what would be the intensity of the particles (number per m^2 sec steradian) arriving within the allowed cone? Lemaître and Vallarta attacked these problems in a series of papers beginning in 1933 and immediately gave a simple answer to the second question: that in any solid angle containing only allowed arrival directions, the intensity would be the same as in an equal solid angle far away from the Earth.

This follows from considering the motion of particles in the six-dimensional phase space $(x_1, x_2, x_3, p_1, p_2, p_3)$. If the equations of motion imply that

$$\sum_{i=1}^{3} \frac{\partial \dot{p}_i}{\partial p_i} + \sum_{i=1}^{3} \frac{\partial \dot{x}_i}{\partial x_i} = 0,$$

which is true in this case, the divergence of the six-dimensional velocity of a point in phase-space is zero, and it follows that the density of points in phase-space, as one follows their motion, is constant. This is Liouville's theorem. Referring to a pencil of rays of velocity v falling in a solid angle $d\Omega$ about the normal to an area dA, a number dN arriving in the time interval dt, the phase-space density is $dN/(dA.v.dt.p^2dp.d\Omega)$, and since p and v are here constant, the intensity $dN/(dA.dt.d\Omega)$ is constant as one follows the particles. (The applicability of the theorem was at first questioned, but later substantiated: the basic equation is in this problem true whether one uses canonical or kinetic momenta.)

The first question cannot be answered by simple arguments: many different trajectories have to be followed by numerical integration, or by simulation with electron-gun models.

The particle motion is best followed in reverse: if ejected from a given point on the Earth in a specified direction, will the particle move away to infinity or not? If the angle lies within Størmer's forbidden cone it will not: it will follow a path forever enclosed in the inner allowed zone of Figure 4c (and will presumably strike the Earth again). However, even if γ is just larger than -1, as in Figure 4b, it is not certain that the particle bouncing about in the inner valley will pass through the jaws to escape: detailed investigation shows that for $\gamma < -0{\cdot}788$ some trapped periodic or quasi-periodic orbits are possible, and no orbit comes in from infinity to end at these particular angles. (Also some arrival directions are excluded because the trajectories would have to pass through the Earth before arriving—there is a "shadow cone" of such directions, at high latitudes, from near the north or south, whichever is the nearer pole.)

The result is that for a given arrival direction (e.g. vertical) at a given location all charged particles with rigidity below the Størmer cut-off are excluded; while in a small interval of rigidity above this there is a "penumbra"—some values are allowed in; above this all rigidities are

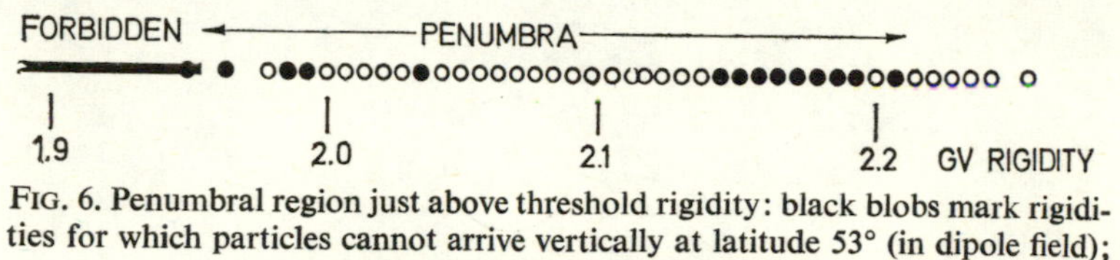

FIG. 6. Penumbral region just above threshold rigidity: black blobs mark rigidities for which particles cannot arrive vertically at latitude 53° (in dipole field); open blobs indicate permitted rigidities.

allowed with undiminished intensity. Figure 6 shows an example of the results when trajectories of particles arriving vertically at latitude 53° were followed numerically: values of rigidity which were found to be "allowed" are marked with open circles, "forbidden" ones by closed circles. Figure 7 shows some of the actual trajectories calculated for particles of different rigidity in the same example (taking a pure dipole field).

Apart from the minor complication of the penumbral effect the implications of the theory were simple. The cosmic rays reaching the Earth include a charged component; and measurement of the cosmic-ray intensity at different latitudes could reveal the momenta (or more strictly ratio of momentum to charge) of the incoming charged component; and the radiation should arrive more from the west or east according to whether the primary particles were mostly positively or negatively charged.

Ironically, it turns out that the aurora cannot be explained in this way.

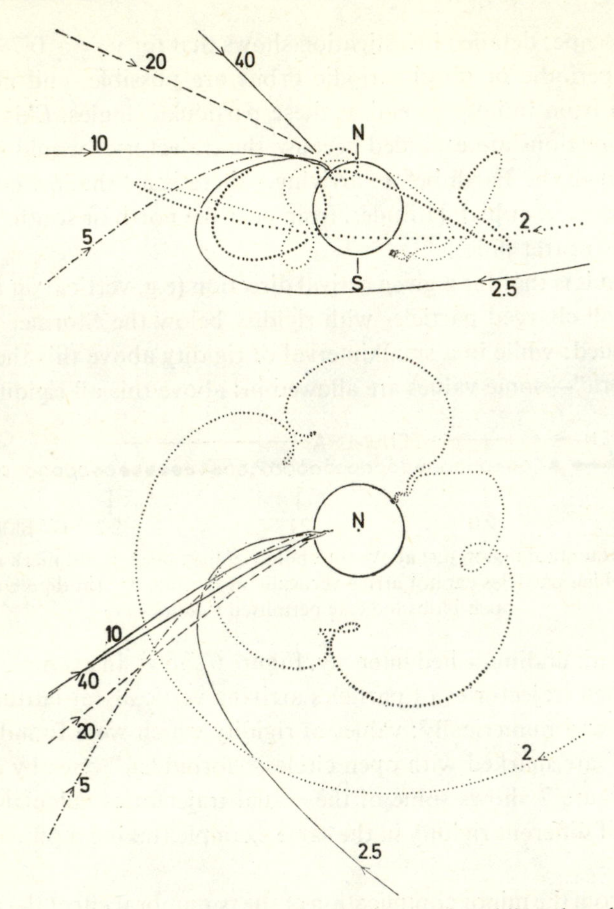

FIG. 7. Projections onto equatorial plane (below) and meridian plane (above) of trajectories (in dipole field) of positively charged particles which arrive vertically at a selected point (53° N). Magnetic rigidities marked in GV.

2.4 Interpretation of the variation of cosmic rays with latitude

Both Clay and Compton found that the intensity of cosmic rays observed near sea level fell by about 16% as one went from latitude 46° (N or S) to the equator. Hence 16% of the radiation observed at high latitudes was apparently traceable back to charged particles reaching the Earth with energies between about 4×10^9 eV and $1{\cdot}5 \times 10^{10}$ eV, which could arrive at that latitude but not at the equator. As for the radiation still present at the equator, it was either more energetic or uncharged.

But the attempts to find an increased intensity in Spitzbergen and Northern Canada had failed; in fact the sharp increase in intensity with increasing magnetic latitude stopped suddenly near 46°: there is a "knee" in the latitude curve.

It might appear then that there are no primary particles of energy below about 3 GeV, to make an appearance at higher latitudes, but Clay (in Paper 3) interpreted the effect as due to the absorption of radiation in the atmosphere: lower energy particles probably did appear, but their effects could not be felt at sea level. If this is indeed the case, these extra particles should be seen at much higher levels in the atmosphere, where the knee should appear at a higher latitude.

(Actually Clay's attempt to estimate the energy lost by a particle on its way through the atmosphere is based on a very incomplete knowledge of particle interactions. The particles which he had measured, inside a thick shield, were not actually primary particles, nor would they generate secondary particles at the rate he quotes. He may not be far out though, in his estimate of the lowest effective energy.)

Since then, much use has been made of geomagnetic cut-off energies to measure the primary energy spectrum (see Chapter 4), but later detailed surveys, using neutron monitors (Rose *et al.*, 1956), and observations in photographic emulsions (Waddington, 1961), have shown that cut-off rigidities predicted from the simple dipole field are not correct. Rothwell and Quenby (1958) found that the cosmic-ray intensity closely followed local anomalies in the magnetic dip angle. For improved calculations of magnetic cut-off rigidities one may consult the tables produced by Quenby and Wenk (1962), who assumed that the entry of a particle "through the jaws" is unaffected by these anomalies, and that it is then guided by the magnetic lines of force, which are distorted from a dipole form, the point of arrival of the particle at the Earth being supposed displaced by just as much as the line of force it attaches to. Shea and Smart (1967) have calculated rigidities by following trajectories by numerical integration. Both used five or six harmonic terms to describe the geomagnetic potential.

The 1930s saw a further pursuit of the cosmic rays into the upper atmosphere, with the last of the manned stratospheric balloon flights, and further development of ingenious automated instruments flown by Regener, Pfotzer, Millikan, Neher and others. Some of the results will be discussed in later chapters. In brief, whilst far more ionization was found at high altitudes, the latitude variation again ceased only a few degrees above the sea-level knee. Curiously, a knee due to limitations of the primary energy spectrum thus appears almost to coincide in position with the knee attributable to atmospheric absorption. Secondly, Pfotzer (1936) made the discovery that the vertical component of the radiation passed through a maximum at a depth $\sim 90\,g\,cm^{-2}$, showing the importance of secondary products, generated from the primary rays above this height. This result derived from the use of a "telescope" of three counters in coincidence, in line, as the effect is largely masked in measurements of the omni-directional flux.

2.5 Evidence for positively charged particles

At Vallarta's invitation, two groups set up counter telescopes in Mexico City. Johnson (1933) ("in a tent on the flat roof of the Hotel Geneve") and Alvarez and Compton (1933) both reported in April 1933 that there was an excess of particles coming from west of the zenith compared with easterly directions: at 45° the difference was about 10%. Rossi obtained a larger result in Eritrea, though earlier measurements at higher latitudes or lower altitudes proved inconclusive. (The atmosphere may clearly blur the alignment of the less energetic particles.) This clearly indicated positively charged particles, and later refinements were to show that few of the particles could be negative. This was an unexpected result, for most of the recognizable particles seen in cloud chambers were electrons, although positrons had made an appearance in the previous year.

Figure 3 shows, in a simple case, how the east–west asymmetry arises.

To summarize the developments described in this chapter: the radiation had been shown to consist in the lower atmosphere of charged particles of an unidentified type, with energies $\gtrsim 10^8$ eV, whilst the primary rays outside the atmosphere included charged particles with energies 3×10^9 to 2×10^{10} eV and above, and the majority of these were positively charged, although physicists were not universally reconciled to this last point for many years. (And Millikan still attributed the equatorial radiation to γ-rays.) Pfotzer's observation showed that the atmospheric particles were probably mostly of secondary origin.

III

Particles Produced by the Cosmic Rays

MOST of the work during the thirties did not at once shed further light on the nature of the primary particles, but was concerned with the link between the primary and the normally observed radiation. The serious application of cloud chambers to the study of cosmic rays transformed the subject and revealed an unsuspected complexity of phenomena: in the end it turned out that the high-energy particles observed in the lower atmosphere are not even secondary, but mostly tertiary products. Gradually a better understanding of particle interactions at high energies developed, alongside the discovery of new types of elementary particles, eventually in embarrassing numbers—the hey-day of cosmic-ray physics.

To make clear the arguments about the nature of these particles, it is helpful to review the interaction of an ionizing particle with the matter it traverses (first, so far as it was established in 1932).

3.1 Ionization effects of fast charged particles

A charged particle moving fast enough through any medium leaves around its path a trail of ions and electrons and excited atoms. The energy expended in ionization and excitation gradually slows down the moving particle and can eventually bring it to rest. The trail also provides the means for detecting the presence of the fast particle. In this section we consider a particle of rest mass M and charge ze moving with velocity v through a medium containing n electrons per unit volume (generally bound in atoms, of atomic number Z, atomic weight A). m is the electron's mass. $\beta = v/c$, and $\gamma = (1-\beta^2)^{-1/2}$. E will denote the total energy of the particle ($E = \gamma Mc^2$), and T its kinetic energy.

The way in which the rate of ionization depends on the velocity and charge of the particle was well known by 1932, from observations on radioactivity, for particle energies up to a few MeV, and since theory accorded well with the variation found, the results could be extrapolated to much higher energies with reasonable confidence. To appreciate this, it is as well to be reminded of the simple physics on which the effect rests. Thus, an electron in the medium at a distance r from the particle's path is affected mainly by the electric Coulomb field of the passing particle, which reaches a peak intensity E_{peak} at closest approach, and is within a factor of two of this for a time $\tau = 2r/v$. If the collision is not almost head-on, the electron may hardly move during this interval, whence the transverse component of the impulse felt by the electron is easily found to be $eE_{\text{peak}}\tau$, the longitudinal component of force reversing and giving very little impulse: so the electron should be ejected with

$$\text{kinetic energy} = \frac{(eE_{\text{peak}}\tau)^2}{2m} = \frac{2e^4}{(4\pi\varepsilon_0)^2 mr^2}\,\frac{z^2}{v^2} \quad \text{(non-relativistic).}$$

In travelling a distance x there will be $2\pi rnxdr$ collisions with impact parameter in the range r to $r+dr$, so the number of electrons ejected in unit length of path with kinetic energy Q to $Q+dQ$ should be

$$dn = \left\{\frac{2\pi e^4 n}{(4\pi\varepsilon_0)^2 m}\,\frac{z^2}{v^2}\right\}\frac{dQ}{Q^2}, \tag{3.1}$$

and the total energy thus lost per unit path length is

$$-\frac{dE}{dx} = \left\{\frac{2\pi e^4 n}{(4\pi\varepsilon_0)^2 m}\,\frac{z^2}{v^2}\right\}\ln\left(\frac{Q_{\text{max}}}{Q_{\text{min}}}\right), \tag{3.2}$$

where Q_{max} is the largest energy which can be imparted (in a head-on collision), and Q_{min} the smallest, which depends on the atomic binding of the electrons. Considering a head-on elastic collision, the conservation laws give

$$Q_{max} = E\left\{\frac{2m\beta^2 E}{(Mc)^2+(mc)^2+2mE}\right\},$$

reducing for $T \ll (M/m)Mc^2$ to $2mv^2\gamma^2[M/(M+m)]^2$, or for $T \gg (M/m)Mc^2$, to E. Though this approximate treatment is clearly invalid for head-on collisions, the same equations, (3.1) onwards, are obtained also when one treats the interaction with each electron as a Rutherford scattering encounter, as is shown in many textbooks. (Even so, (3.1) is not quite correct for the very rare collisions in which a very large amount of energy is transferred to one knock-on electron, for this depends on the structure of the particle, and Q becomes relativistic: for example for a particle obeying Dirac's equation, but more massive than an electron, the expression (3.1) has to be multiplied by the factor $(1-\beta^2 Q/Q_{max})$. This is of little consequence, however; and in such close encounters one might anticipate other processes.) The value of Q_{min} is less easy to decide, as this rough treatment is inadequate for the more distant collisions, and a calculation is required which takes account of atomic binding.

Following calculations by Bohr, Bethe and Bloch, the expression for energy loss comes to

$$-\frac{dE}{dx} = K\frac{z^2}{\beta^2}\left[\ln\left(\frac{Q_{max}}{Q_0}\right)-\beta^2+\ln\left(\frac{2mc^2 Q_0\beta^2\gamma^2}{I^2}\right)-\beta^2\right], \quad (3.3)$$

for particles appreciably more massive than electrons (using the corrected form of (3.1)). I is the mean ionization potential for the various bound electrons, roughly given by (12·8 eV)Z. The first two terms in the square brackets relate to close collisions in which electrons are ejected with energies greater than some arbitrary value Q_0, whilst the last two terms refer to smaller energy transfers. Here, $K = 2\pi e^4 n/[(4\pi\varepsilon_0)^2 mc^2]$. Provided $T \ll (M/m)Mc^2$, and measuring path length ds (= density.dx) in g cm^{-2},

$$-\frac{dE}{ds} = 0{\cdot}154\left(\frac{2Z}{A}\right)\frac{z^2}{\beta^2}\,[11{\cdot}29+2\ln(\beta\gamma)-\beta^2-\ln Z]\ \text{MeV/(g cm}^{-2}). \quad (3.4)$$

For fast electrons, the situation is somewhat different, with identical particles colliding, and the result is:

$$-\frac{dE}{dx} = \frac{K}{\beta^2}\left[\ln\left(\frac{mc^2\beta^2\gamma^2T}{2I^2}\right)-\left(\frac{2}{\gamma}-\frac{1}{\gamma^2}\right)\ln 2+\frac{1}{\gamma^2}+\frac{1}{8}\left(1-\frac{1}{\gamma}\right)^2\right]. \tag{3.5}$$

This varies with velocity in a similar way to (3.3), but is rather less.

For the purpose of observing charged particles, the important points about the energy loss are that (a) the number of ions produced along or near the path is proportional to the energy lost by the particle, and (b) the rate of energy loss (and hence ionization) may be written as

$$I = z^2f(v), \quad \text{or} \quad I = z^2F(\gamma); \tag{3.6}$$

thus it does not depend appreciably on the mass of the particle, given v.

The function F is plotted in Figure 8: the ionization is very large for low velocities (ignoring velocities too low to ionize), falling roughly as $v^{-1\cdot4}$ to a minimum near $v = 0{\cdot}97c$, and then rising very slowly at ultra-relativistic velocities. The minimum energy loss is about 1·8 MeV per g cm^{-2} for air, falling to 1·1 MeV per g cm^{-2} in the heaviest elements (but rather less for electrons). In 1932 the range $\gamma \gg 1$ had not been explored, and the only important modification introduced since then is the recognition that the ionization rises rather little in this range, particularly in dense media. (The figure includes this effect.)

So provided the charge of the fast particle is known—it will usually be of the same magnitude as the electron's charge—the ionization is a measure of the particle's velocity, and so, unless $v \doteqdot c$, it may be possible to deduce the mass M of the particle if either its momentum, total energy or range can also be found. Momentum is often measured by observing the curvature of the trajectory in a magnetic field, for projecting this onto a plane normal to the field lines gives a circle of radius r, the component of momentum p in this plane being given by the relation:

$$cp = 300\ Brz\ \text{eV (units: gauss, cm)} \tag{3.7}$$

—in the form often quoted. (Or: $cp = 0{\cdot}3\ Brz$ GeV, using Wm^{-2}, m.)

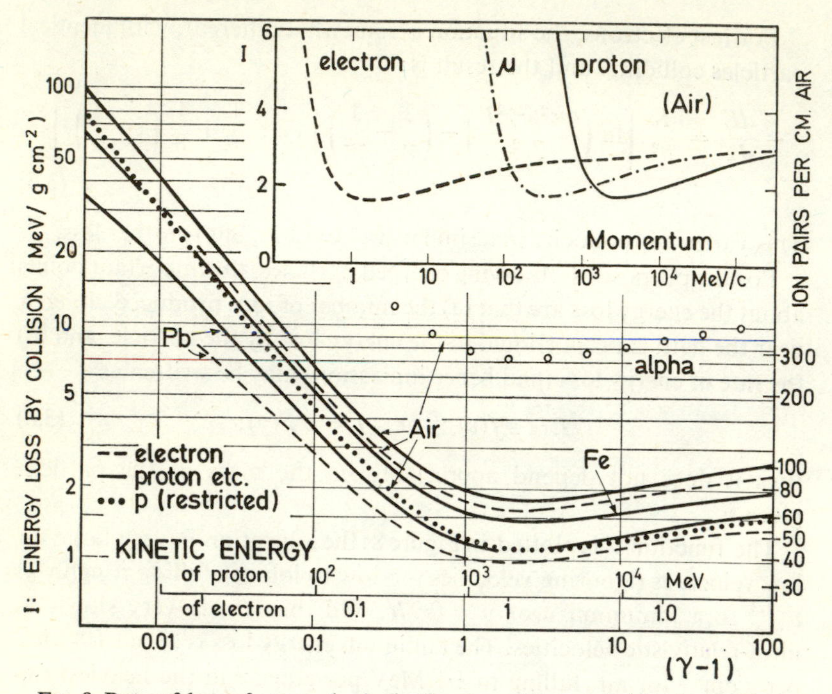

FIG. 8. Rate of loss of energy by ionization in various materials. $(\gamma-1)$ is the ratio of kinetic energy to rest energy: actual kinetic energies are also given for protons and electrons. Inset shows energy loss (in air) as function of momentum, which is often measurable.

Except as above, the trajectories will be nearly straight lines for particles heavier than electrons (or for electrons of many MeV), and these energy losses will eventually stop the particle after a distance R, known as the range, which may be calculated by integrating the expression for energy loss, and is shown in Figure 9 as a function of momentum. For given v, it follows from (3.6) that $R \propto M/z^2$, so R varies strikingly with M.

The trail of ionization left by the particle, and made visible in cloud chambers, and later in emulsions, show some structure. One may distinguish (i) ejection of electrons with energies up to some limit Q_0, say about 1 keV, and (ii) ejection of more energetic electrons, which

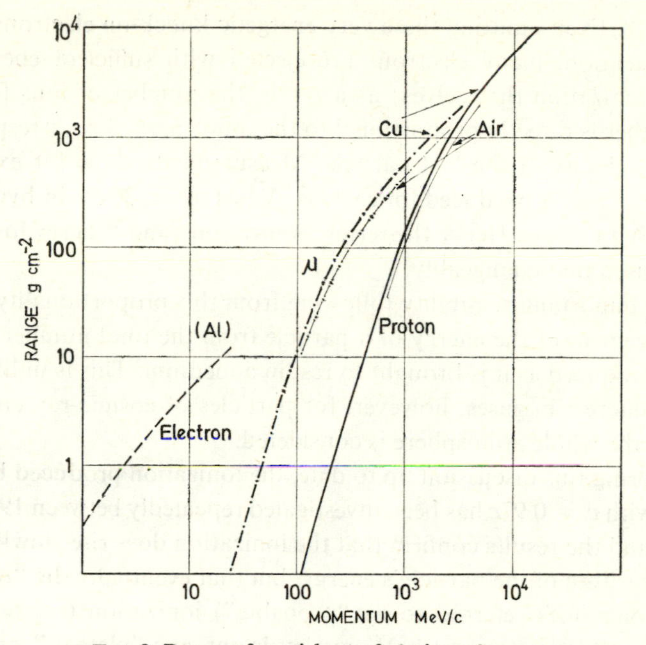

FIG. 9. Range of particles as function of energy.

have enough energy themselves to make very dense local clusters of secondary ionization. The more energetic ones will form distinct branch-tracks of their own, known as δ-rays, and from equation (3.1) the number with energy greater than Q per g cm^{-2} of track will be

$$n_\delta(Q) = \frac{2Z}{A}\,\frac{0{\cdot}077\ \text{MeV}}{Q}\,\frac{z^2}{\beta^2} \qquad (\text{if } Q \ll Q_{\text{max}}), \qquad (3.8)$$

a relation which has been employed when the ionization is very dense, as in the study of highly-charged particles.

Because the presence of one or two very energetic δ-rays can cause large upward fluctuations in the energy loss in a very thin layer of material, measurements of the "restricted ionization", counting only ionization under heading (i), is often preferred when making visual observations. Its magnitude may be calculated from equation (3.3) by omitting the first two terms in the square bracket, and is indicated in

Figure 8. Even omitting these very energetic knock-on electrons from our attention, many electrons are ejected with sufficient energy to cause ionization themselves: as a result, the number of ions formed altogether is directly proportional to the total energy loss, irrespective of the velocity of the fast particle. Measurements show for example that 1 ion pair is produced for each 34 eV lost in air, 36 eV in hydrogen or 26 eV in argon. Hence the terms "ionization" and "energy loss" are often used interchangeably.

One important possibility following from this proportionality is the measurement of the energy of a particle from the total number of ion pairs produced as it is brought to rest in a medium. This is unlikely to be of interest in gases, however, for particles of cosmic-ray energies, unless the whole atmosphere is considered.

To bring this discussion up to date, the ionization produced by particles with $v > 0{\cdot}97c$ has been investigated repeatedly between 1938 and 1960, and the results confirm that the ionization does rise slowly, with the logarithm of the particle's energy, but that eventually the "restricted" (sometimes referred to as "probable") ionization (i.e. ignoring large energy transfers >1 keV, say) levels out, at a "plateau", an effect first anticipated by Swann (1938) and Fermi (1940). The rise is associated with a gradually increasing radius at which ionization and excitation can take place, but in condensed media this range is limited by the electrostatic screening effect of nearby electrons. In solids the rise is thus relatively small, the grain density in nuclear emulsions being found to rise only 13% above the minimum, but in gases at atmospheric pressure the rise may be about 40%, whilst 70% may be possible at 0·1 atm. (The total energy loss does, formally, continue to rise more slowly with increasing energy, but only through the increasing Q_{max} transferred in exceptionally rare collisions.) Sternheimer (1961) gives a more detailed discussion.

Roughly speaking, most cosmic-ray particles have velocities $>0{\cdot}9c$ and so have rather similar ionization densities, close to the minimum. An electron would stop within 0·02 g cm^{-2} of the point where it is slow enough to give ionization twice the minimum value: the strongly ionizing part of a proton's track would be 1800 times as long.

3.2 Cloud chamber tracks of cosmic-ray particles

It was not long after Skobelzyn had reported his discovery of cosmic-ray tracks in a cloud chamber before other groups began observations. In particular, Anderson and Millikan at Caltech, Kunze at Rostock and Blackett and Occhialini in Cambridge used cloud chambers combined with large electromagnets to observe magnetic deflections. The earlier workers relied on random expansions to give perhaps one good track in 50 photographs, but the latter pair developed a system for expanding the chamber within 1/100 sec of Geiger–Müller counters detecting a particle entering and leaving the chamber, so that tracks were always seen (though not a random sample). These early chambers were about 6 in. in diameter, but extremely shallow. Anderson and Kunze employed fields up to $\sim$ 20 kilogauss, and the maximum radii of track curvature measurable before track distortion became serious corresponded to momenta of $\sim$ 4 GeV/c (i.e. $cp = 4 \times 10^9$ eV); and at once they found energies of $\sim 10^9$ eV to be typical, though some particles were too energetic to be measurably deflected. The deduction of similar energies from the geomagnetic effects followed in the same year.

Quite unexpectedly, there were as many positively as negatively charged particles; and groups or "showers" of particles were quite frequent, seeming to originate in the massive windings of the magnet, probability arguments showing that they could not be associated by chance. Though the first natural presumption was that the positive particles were protons, the ionization did not look right. In order to study their production and help identify them, Anderson put a lead plate in the cloud chamber, for the particles to pass through. Anderson (1932) and Blackett and Occhialini (1933) then came to the same conclusion: The positive particles could not be protons because they showed low specific ionization and long range even at momenta below 300 MeV/c (cf. Figure 8); in fact for a few tracks below 100 MeV/c the combination of momentum and ionization necessitated a mass closer to that of the electron.

In a paper reprinted here as Paper 4, Anderson published examples of a few selected photographs which made this conclusion hard to avoid: the particles were positive electrons, or positrons. It is not possible from these pictures to prove the mass of the particles equal to that of an electron; but it is certainly much less than that of a proton, or the energy losses in the plate, and the ionization, would have been higher: nevertheless, the energy losses in the lead plate for positrons and electrons were appreciably greater than the 18 MeV or so per cm which might have been expected from equation (3.4) (and for Anderson's celebrated Figure 1 the loss is close to the expectation for a muon, but then the ionization would have been 5 times higher above the plate than below). Anderson thought 20 m_e an upper mass limit: the reader may like to make an estimate by comparing the ionization with that of a fast electron in his Figure 3 ($< 30m_e$?). Only for the particles of lowest momentum could this identification be proved; and many of the positive particles were probably not positrons as supposed.

The origin of the positives was the next question, and to Blackett and Occhialini (1933) this seemed to be in the complex showers of particles in which they observed tracks radiating from a region close above the chamber or in a 4 mm lead plate. Frequently, when a shower entered from above, a further multiplication of tracks occurred in the lead plate.

Concerning showers, the situation appeared thus to these authors:

The mean free path for knocking out energetic δ-rays could be calculated (e.g. by equation 3.8), so "...from this figure, the chance of two or more secondaries being produced independently nearly at the same point is very small. So when one track is seen to branch at one point into three or more tracks, it is impossible to resist the conclusion that some nuclear interaction has taken place. If an incident photon is considered the same argument is valid.

"...There are three possible hypotheses that can be made about the origin of these particles. They may have existed previously in the struck nucleus, or they may have existed in the incident particle, or they may have been created during the process of collision. Failing any independent evidence that they existed as separate particles previously, it is reasonable to adopt the last hypothesis....

"On the view that the neutron is an indivisible particle and that there are no free negative electrons in light nuclei, it follows that both the negative and positive electrons in the showers must be said to have been created during the process. If, however, conservation of electric charge is to be fulfilled, then positive and negative electrons must be produced in equal numbers, for there can hardly be many protons created on account of the large energy ($E \sim Mc^2 = 940$ million volts), required to create one.

"In this way one can imagine that negative and positive electrons may be born in pairs during the disintegration of light nuclei. If the mass of the positive electron is the same as that of the negative electron, such a twin birth requires an energy of $2mc^2 \sim 1$ million volts, that is much less than the translationary energy with which they appear in general in the showers."

This interpretation in outline was readily accepted, for the theoretical framework for describing the processes observed was already available in Dirac's theory of the electron, from which positive electrons were expected. At the time the theory was in some difficulty in explaining the asymmetry in mass between the only known positively and negatively charged particles, but the appearance of the positron instilled confidence in Dirac's theory, which had successfully accounted for the magnetic moment of the electron and the fine structure of the energy levels of the hydrogen atom, so that in the next few years its implications regarding the interactions between electrons and γ-ray photons were worked out, largely by Heitler, Bethe, Oppenheimer and their collaborators, with great success. The predicted lifetime $\sim 10^{-9}$ s before annihilation of slow positrons by electrons gave a good reason why positrons had not been observed before.

The theory did not, however, predict the appearance of electron-positron pairs when a nucleus was disrupted, but rather explained the production of a single pair when a γ-ray of energy $> 2mc^2$ passed through the electric field close to a nucleus, or, much more rarely, if an electron passed close to a nucleus. This basic process of pair production explained an anomaly in the absorption of energetic γ-rays (> 2 MeV) particularly by heavy elements: the increased absorption was due to the new process taking place, and the positrons produced were later annihilated, radiating pairs of quanta of energy mc^2 (0·51 MeV), which had also been observed. Immediately the idea that a very high energy γ-radiation was very penetrating was overthrown. Soon the same happened to electrons, and there arose a problem of explaining how cosmic rays penetrated to the bottom of the atmosphere, and further.

The results of these calculations with the quantum electrodynamics are essential to an understanding of many cosmic-ray phenomena.

3.3 The interaction of high-energy electrons and photons

(a) RADIATION BY HIGH-ENERGY ELECTRONS

As already remarked, Anderson found that electrons and positrons lost more energy than one would expect on the basis of section 3.1, when they passed through a lead plate, and Heitler and Sauter (1933) pointed out that an electron with an energy of a few MeV or more should lose energy more rapidly than previously assumed, by radiating energetic photons from time to time, which they referred to elsewhere as bremsstrahlung (= "braking radiation"), emitted when the electron (or positron) is suddenly deflected by passing close to a nucleus. Classical electromagnetic theory could be used to predict a radiation intensity proportional to the square of the particle's acceleration, i.e. $\propto z^2Z^2/M^2$, with notation as before; so the effect is most marked for light particles (not protons) passing close to a nucleus of high charge (e.g. Pb). Quantum electrodynamics retains the same proportionality in the energy radiated, and thus in the probability that a photon will be emitted. (Deflection by one of the atomic electrons can also cause radiation, and alters the Z^2 dependence to approximately $Z(Z+1{\cdot}3)$.)

If there are n nuclei per unit volume, the average rate of loss of energy by an electron of energy E may be written as

$$-\left(\frac{dE}{dx}\right)_{\text{rad}} = nE\Phi(E),$$

where $\Phi(E)$ rises with energy initially and then becomes constant when the atomic electrons have an important effect in screening the nuclear charge. At very high energy,

$$\Phi = 4Z(Z+1\cdot3)\alpha r_0^2 \left\{ \ln\left(\frac{183}{Z^{1/3}}\right) - \left(\frac{Z}{152}\right)^{7/4} \right\}, \qquad (3.9)$$

where r_0 is the classical electron radius, $e^2/(4\pi\varepsilon_0 mc^2)$, and $\alpha = 1/137$. (The term in $(Z/152)$ represents approximately corrections proposed 20 years later to the earlier value calculated with the Born approximation, but it changes the 1934 (Bethe and Heitler) result by less than 10%.)

It is convenient to define a unit of length, the radiation length, x_0, to give a scale to this process—a characteristic decay length for the energy of a very energetic electron:

$$-\frac{dE}{E} = \frac{dx}{x_0},$$

considering only losess by radiation. From equation (3.9) one finds the radiation length,

$$x_0 = \frac{1}{n\Phi(\text{large } E)}$$

$$= \frac{717A}{Z(Z+1\cdot3)\{\ln(183/Z^{1/3}) - (Z/152)^{7/4}\}} \text{ g cm}^{-2} \qquad (3.10)$$

for a material of atomic weight A and atomic number Z.

Figure 10 shows how the rate of energy loss will vary with the electron's energy.

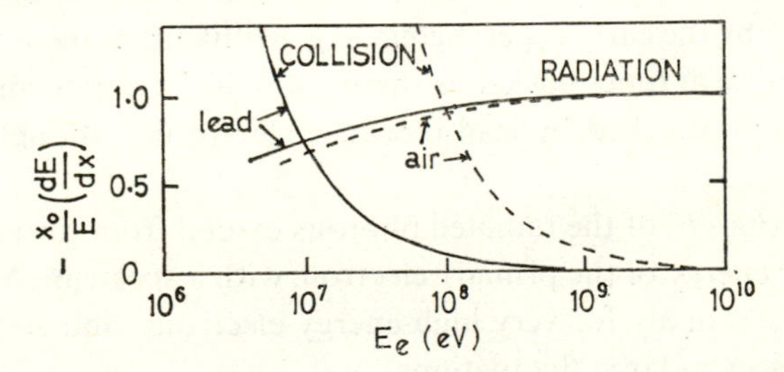

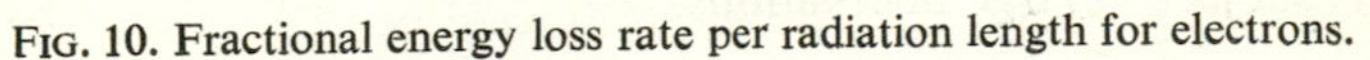
FIG. 10. Fractional energy loss rate per radiation length for electrons.

The other parameter which characterizes the energy domain in which radiation phenomena dominate is the critical energy, ε, which is the energy lost by ionization (and excitation) while traversing one radiation length (evaluated for $E = \varepsilon$): radiation rapidly becomes the dominant energy loss above this energy.

The rapidity and the energy domain of the loss by radiation may be judged from the values of x_0 and ε given in Table 3.1.

TABLE 3.1.
TYPICAL VALUES OF RADIATION LENGTH AND CRITICAL ENERGY

Substance	Radiation length	Critical energy (MeV)
Hydrogen	60 g cm^{-2} = 6·7 km	335
Air	36·1 g cm^{-2} = 280 m	87
Water	35·5 g cm^{-2} = 35·5 cm	86
Iron	13·7 g cm^{-2} = 1·76 cm	21·6
Lead	6·3 g cm^{-2} = 5·6 mm	7·3

Roughly, $\varepsilon = 600$ MeV/Z for condensed materials, though values quoted by different authors differ appreciably.

So for a large proportion of cosmic-ray electrons and positrons radiation gives the dominant energy loss; and since the vertical depth of the atmosphere at sea level amounts to 28·4 radiation lengths, no primary electrons would be expected to penetrate it. Another noteworthy point is that the effectiveness of each atom in slowing an electron is proportional to Z^2, not to Z as in the case of ionization: lead is thus much more effective than the same mass of a light element. This was not anticipated by the early experimeters (e.g. Millikan, Bothe and Kolhörster) who were at times misled, as there is a "soft" component of cosmic rays, readily absorbed by lead (because electronic), though yet quite energetic.

The energies E' of the radiated photons extend from zero right up to the kinetic energy of the primary electron, with a spectrum $N(E')dE' \propto dE'/E'$ very roughly for very high-energy electrons, and so the energy loss is subject to large fluctuations.

(b) BEHAVIOUR OF HIGH-ENERGY PHOTONS

Immediately after the discovery of the positron, it was found that for γ-rays of many MeV energy their absorption to form electron–positron pairs was much more important than Compton scattering as a loss process. Experimentally, the pair-producing radiation generated pairs in the first few mm of lead, and rapidly disappeared. Theoretically, this behaviour could be expected for γ-ray photons: Oppenheimer and Plesset (1933) made a preliminary calculation on the basis of Dirac's theory, as did Heitler and Sauter: then Bethe and Heitler (1934) made detailed calculations on bremsstrahlung and pair production, and predicted a mean free path for pair production very close to 9/7 radiation length for very energetic photons—loss would continue to rise with energy if the atomic electrons did not shield the nuclear charge. (The variation of the pair-production rate with energy is shown in Figure 11,

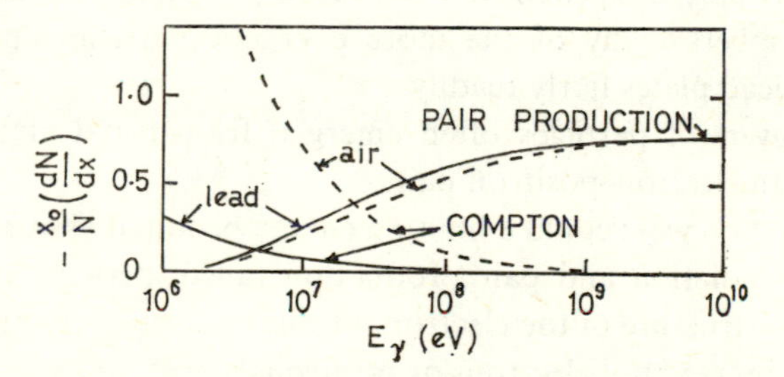

FIG. 11. Fractional loss rate per radiation length from a beam of photons.

and compared with the absorption due to Compton scattering.) In the air this free path is 47 g cm^{-2}, or less than one-twentieth of the depth of the atmosphere: thus γ-rays found in the lower atmosphere must be of secondary origin.

As the radiation length is only a few mm in lead, bremsstrahlung or pair production is very likely in the thickness of the plates used in the cloud chamber experiments. It had indeed been remarked that when a group of particles entered a chamber, further electrons and positrons were frequently generated in the lead plate, by a non-ionizing ray.

3.4 The penetrating particles

In 1934, the correctness of the theory of quantum electrodynamics was doubted, however, on account of the following difficulties:

(a) Whereas theory predicted that bremsstrahlung should reduce the energy of an electron (or positron) by a factor e in 36 cm of water (or an equivalent mass of air), and the mean free path of a high-energy photon should not be much different, cosmic rays were evidently capable of very much slower attenuation, both as regards the general intensity of radiation, which Regener (1931) had traced to a depth of 237 m in Lake Constance, by lowering a self-recording electrometer, where the attenuation length $I/(dI/dx)$ increased continuously, exceeding 50 m water at 100 m, and also as regards individual particles. For the results of Bothe and Kolhörster now appear in a new light, as their gold absorber amounted to 12·3 radiation lengths, and according to Rossi many particles penetrate 1 m of lead (175 radiation lengths!). In cloud chambers many of the more energetic particles appeared to penetrate lead plates fairly readily.

(b) Showers of particles often emerged from metal plates, rather than just an electron–positron pair.

The question was: could a limiting energy be found above which the theory of radiation and pair production failed, perhaps because of some finite structure of the electron, or alternatively were many of the particles observed not electrons or positrons?

Anderson continued his important measurements on energy losses in a lead plate, while Bethe and Heitler made the detailed calculations on radiation processes whose results were stated in the previous section. With the effect of screening of the nuclear field allowed for, many but not all of the lower energy particles were found to behave satisfactorily, but at high energies the predictions were clear, and in conflict with (a). Weiszäcker (1934) and Williams (1934) showed that the process of radiation could be regarded as a scattering by a fast electron of Fourier components of the nuclear electric field which it saw moving past, and that the most important wavelengths were not unduly short, so that there was no reason to suppose that a finite electron radius could be relevant or the theory of Compton scattering would be wrong. Williams suggested that the energetic particles must be protons or hypothetical negative protons.

Furthermore, there were showers. Although some showers in which several particles diverged from a point pointed to nuclear disintegrations, most showers showed a more gradual growth as they progressed, for which many people suggested an explanation: fast electrons radiated γ-rays, which produced further electron–positron pairs, and so on, to build up an electron–photon cascade, all particles moving in much the same direction. The effectiveness of different atoms in developing small showers varied as Z^2, as for the radiation processes.

It was in this atmosphere of doubt that Bhabha and Heitler, and independently Carlson and Oppenheimer, approached the theory of the development of electron–photon cascades, as may be read in Paper 5, "On multiplicative showers", by the latter authors, part of which is reprinted in Part 2. The authors were evidently rather surprised to find the phenomenon of shower production to be in favour of the theory, notably at high energies.

Bhabha and Heitler (1937) computed the growth and decay of the first few generations of particles formed in this way to show that the growth of showers was correctly described, but this approach soon became tedious, and Carlson and Oppenheimer (1937) took a complementary approach in which a large cascade is treated as an almost continuous distribution of particles in energy and space, developing in a manner rather like diffusion. Their conclusions are reprinted as Paper 5, omitting the long mathematical details in which they made an approximation which is hard to assess (and was later proved unnecessary in further work by Landau and Rumer (1938) and Snyder (1938)). Their simplified explanatory description of the shower as a successive doubling, and then absorption, helps one to understand the result that the number of particles builds up to a maximum at a distance of very nearly $\ln(E/\varepsilon)$ radiation lengths, where it has about E/ε particles (E being the energy of the initial particle), before dying away. The distance at which the maximum number of particles should be found was not only in rough agreement with experimental observations on showers, but it also appeared to explain the rise in intensity of the cosmic rays on entering the atmosphere, the depth and magnitude of the Pfotzer maximum being quite reasonable if the incoming particles were electrons (although, as will appear, this latter remark started a false trail). Bhabha and Heitler were at first satisfied with this cascade phenomenon as a possible explanation of penetration of particles to sea level, but Carlson and Oppenheimer, whose method of calculation was more appropriate to the tail end of a shower, found that although it seemed one had mainly an electron–photon component at high altitude, there must be another component of cosmic rays, much more slowly absorbed. They mention in their paper the possibility of a new type of particle.

Meanwhile, the suspicion that other particles were present was growing, and in 1937 the answer appeared in four cloud chambers. The main question was resolved when Neddermeyer and Anderson showed clearly (in Paper 6 reprinted in Part 2) that the unusual penetrating power was not simply a function of the energy of the particle: some energetic particles did behave as predicted, and radiate readily—and these appeared as members of showers: they were clearly electrons. But many particles of a similar energy behaved differently: they arrived alone, and showed no sign of radiating on passing through a platinum plate, so they were evidently much more massive than electrons, but less massive than protons as they did not ionize heavily even when their momentum was less than 500 MeV/c.

From Tokyo, Nishina, Takeuchi and Ichimiya (1937), using a larger cloud chamber and strong magnetic field, reported the same conclusions. They, and also Street and Stevenson at Harvard, set out to deduce the mass of the penetrating particles by simultaneous measurement of the momentum (from curvature) and rate of energy loss of the particles, assuming they lost energy only by the usual ionization and excitation processes. Most particles showed near-minimum ionization but a few were travelling sufficiently slowly to give a rate of energy-loss on the mass-sensitive part of the curve (see inset, Figure 8 on p. 30), and the Tokyo group reported the mass of one particle to be 1/7 to 1/10 that of the proton. Street and Stevenson (1937) produced an example estimated at 130 times the electron mass. (The method used by the latter was rather difficult, as to measure the ionization accurately the droplets were counted: this required some diffusion of the track ions to be permitted, which made the track curvature less well defined.) In Paris, Leprince-Ringuet and Crussard found evidence for a particle of intermediate mass, for instance from the detailed kinematics of an unusual collision with an electron.

The new particle was usually referred to as the mesotron in literature of the period: it later became "meson", and now "muon". Its mass became better defined, the eventual precision attainable with cosmic-ray range-momentum measurements (Brode *et al.*, 1950) being $206 \pm 2m_e$. The accepted value is 206·8. The Tokyo group reported that 80% or more of the more energetic particles at sea level were not electrons, and presumably mesons.

The less penetrating components of cosmic rays, then, have been filtered out before reaching sea level; what remain are these mesons, which lose energy almost exclusively by ionization (i.e. about 2×10^9 eV in the whole of the atmosphere). The radiation length of a meson should be $(207)^2$ times that of an electron, so that radiation should predominate as an energy loss only above $3{\cdot}7 \times 10^{12}$ eV in air.

3.5 The origin of the mesons

The mesons were not the primary cosmic-ray particles, for several arguments and eventually direct cloud-chamber and counter observation showed that the meson was unstable, decaying into an electron and other particles (neutrino and antineutrino in fact) with a mean proper lifetime of 2·2 μs: thus they must have been produced in the atmosphere, and those electrons found near sea level in defiance of the theory of absorption could largely have been produced locally by meson decay (and as δ-rays).

A detailed discussion of this instability is not appropriate here, but a few points are of interest in passing. The indirect argument for the instability arose from careful observations of the altitude dependence of the penetrating particles, which showed that more particles disappeared in a given number of g cm^{-2} of air than in the same mass of water or other condensed light material, contrary to the original observation which persuaded Millikan 12 years previously that there was a cosmic radiation. Guided by a theoretical idea of Yukawa (1935) that there should be a particle of mesonic mass which decayed into an electron and neutrino (being supposed responsible for nuclear β-decay), Kulenkampff (1938) proposed such a decay to account for the absorption anomaly, for in addition to absorption by ionization, the particles might suffer decay whilst traversing a long path in the air, but hardly whilst passing through a few g cm^{-2} of a solid material. A temperature dependence of the cosmic-ray intensity also came to light, and this was attributed to the same cause (Blackett, 1938), for a warmer atmosphere is less dense, and mesons have to travel further to traverse the same mass of air, so that more of them can decay en route. (It turned out that about a third of the latitude variation of radiation seen at sea level is due to the fact that the equatorial atmosphere is hotter.) And in the course of the ensuing investigations the relativistic effect of time dilation showed up clearly: the mean apparent lifetime of a meson of total energy E, and rest mass M, is $(E/Mc^2)\times 2{\cdot}2$ μs. (Mesons of 10^9 eV kinetic energy, for example, have thus a mean life of 23 μs and 13% would decay on a flight of 1 km through the air, as against 78% if the lifetime remained 2·2 μs.)

In the late thirties there was some support for a dualistic theory of cosmic rays, with the electronic and penetrating components originating from different primary particles, and the early cascade calculations (e.g. Carlson and Oppenheimer) favouring primary electrons (or positrons) to generate the secondary flux. However, increasing numbers of unmanned balloon flights, at various latitudes, yielded less satisfactory results, showing for instance that the initial growth of secondary particles near the top of the atmosphere did not look like a rapid successive-doubling process that characterizes the start of a cascade. However, what was generally regarded as decisive evidence of the secondary nature of both the meson and the electronic components came from the report by Schein, Jesse and Wollan on a series of experiments with shielded Geiger counters carried by balloons to very high altitudes. This letter on "The nature of the primary cosmic radiation and the origin of the mesotron" is reprinted as Paper 8.

In this paper Schein *et al.* report that they measured the number of particles at various altitudes which would penetrate various thicknesses of lead, from 7 to 33 radiation lengths, without producing showers. The important result was that this "hard" radiation, which could not include appreciable numbers of electrons, rose steadily with increasing altitude until it appeared to account for the whole of the radiation at the top of the atmosphere. Their observation that on entering the atmosphere the number of penetrating particles does not at first increase as might be expected if secondaries are being generated is a little surprising, but apparently not of great significance: it seems that many low-energy primaries are present, whose disappearance balances the generation of secondaries; and at lower geomagnetic latitudes an initial increase has been found. However, the authors could conclude that very few electrons of energy between the geomagnetic threshold and 10^{12} eV were present in the upper atmosphere. The observations, moreover, were made within 1 radiation length of the top of the atmosphere, giving no opportunity for primary electrons to vanish. Hence the intense soft radiation around the Pfotzer maximum must comprise low energy secondary electrons, which could quite possibly arise from meson decay. The primary particles were in all probability protons, to agree with the implications of the east–west asymmetry.

Swann (1941) had been advocating a similar view for several years, having been impressed in particular by the near isotropy of the electrons near the Pfotzer maximum, which his collaborators found in manned balloon flights in 1936, and suggesting to him that the decay of slow mesons was their origin.

This is nearly our view today, except that some of the more important decaying mesons are π°-mesons.

These balloon experiments reported in Paper 8 were probably the first to observe the primary particles directly whilst distinguishing crudely between different types of particle (though not between protons and mesons). And yet there was one more link in the chain!

3.6 π- and μ-mesons

The discovery of a meson had been welcome in the context of nuclear physics, since Yukawa's discussion of nuclear forces had suggested that the quanta of the nuclear force field should appear as particles about 300 times as massive as electrons. However, it was soon apparent that the meson's properties were not those of the quantum of the nuclear field: again, the difficulty was that the cosmic-ray particles lose too little energy in passing through matter. Clearly these mesons did not interact strongly with atomic nuclei. For one thing, a meson must pass through ~ 12 atomic nuclei in traversing the atmosphere, and so the effect of any one encounter must be negligibly small; and in the laboratory, counter experiments on stopped mesons by Conversi, Pancini and Piccioni and also by Sigurgeirsson and Yamakawa in 1947 showed that even when captured into a Bohr orbit a negatively charged cosmic-ray meson could exist for a relatively long time inside the nucleus. (Later analysis showed that the bound meson had a lifetime of $\sim 5 \times 10^{-9}$ s inside nuclear matter, a factor of 10^{13} larger than expected for a Yukawa particle.)

Although a few theoreticians made some very shrewd guesses about what was happening, the elucidation of the interactions between mesons and nuclei came with the application of a relatively new observational technique—the nuclear photographic emulsion.

The fact that the tracks of heavily ionizing particles in photographic emulsions could be developed as lines of silver grains was known since 1900, and nuclear disintegrations caused by cosmic rays had been detected in 1937, but this observational technique lagged far behind cloud chambers until the middle 1940s, when it was further improved for studying nuclear reactions and was then vigorously applied to the study of nuclear interactions of cosmic rays, particularly by Powell and his collaborators at Bristol, in conjunction with the Ilford company, who evolved methods of preparing and developing special emulsions, which were very much thicker and richer in silver halide than normal photographic emulsions. A sufficient amount of ionization in a silver halide crystal can render it developable, and tracks are formed as strings of developed grains, the number of grains (or grain density) along a track being proportional to the rate of energy loss by the particle, though it also depends on the details of development and on the time elapsing before development. The nuclear emulsion makes visible even more detail of particle interactions than does the cloud chamber, showing, in effect, what happens inside the plates of a cloud chamber, but even at the time of the method's great triumph in the discovery of the π-mesons only densely ionizing tracks, formed by singly-charged particles with a velocity $< 0{\cdot}4c$ (or by multiply charged particles), could be detected. As a result, in emulsions left for a period at high altitude where the cosmic-ray intensity was high, tracks of particles could be seen emanating from nuclear disintegrations but there was usually no trace of the particle which struck the nucleus, which was presumably relativistic. However, Perkins (1947) observed one disintegration evidently caused by a slow charged particle, in which the only apparent source of the energy release of 100 MeV was absorption of the rest mass energy of the particle. The density of grains and amount of scattering of the track were consistent with a meson mass. This showed that mesons could be violently absorbed by nuclei; soon it was seen that mesons were among the particles emitted in nuclear interactions.

Long balloon flights having always been hazardous in the British Isles, the Bristol group arranged for emulsions to be exposed for a few months in the mountain laboratories which were becoming used for cosmic ray work, at Pic du Midi and Chacaltaya (Bolivia). Lattes, Muirhead, Occhialini and Powell (1947) then reported finding the tracks of many slow mesons which had been brought to rest in the emulsions, and in two cases a second meson track was seen to start from where the first one stopped. They could not satisfactorily explain this through energy gained in some nuclear interaction, and were left with the conclusion that one kind of meson was formed which decayed into another lighter one. If the latter was the well-known cosmic-ray meson, which itself decayed into an electron, the electron would be relativistic, and its track could not be seen.

Particles were identified as mesons either by range-grain density or range-total number of grains relationships, analogous to the range-velocity or range-energy relations, or from the multiple Rutherford scattering visible in the track of the particle, for at a given velocity (given grain density) the r.m.s. scattering angle per unit path length is inversely proportional to the particle's rest mass.

Good evidence that a decay process was involved was presented in the following paper on "Observations on the tracks of slow mesons in photographic emulsions", reprinted as Paper 9 in Part 2, from which some idea of the method of investigation may be obtained. Here, incidentally, Lattes, Occhialini and Powell introduce a terminology for mesons behaving in different ways, some of which has stuck: the π-meson for the primary particle in the decays, and μ for the secondary meson, which they identified with the penetrating meson observed at sea level. They claimed that they saw positively charged particles decaying, for the negative ones were attracted to nuclei as they stopped, and rapidly absorbed. From the important observation that μ-mesons of unique energy are generated in the decay of π-mesons, it is clear that one other, neutral, particle is also produced. It seems that the two decays analysed in detail were rather misleading, as the authors obtained too high a value for the mass ratio m_π/m_μ, and hence preferred the wrong alternative of a massive neutral particle: later work favoured a (massless) neutrino, and a π-meson rest mass of $273{\cdot}2m_e$. Later, in improved emulsions, μ-decay could also be seen.

The first paper on π-μ decay was published at a very opportune moment, for it was received at a conference at Shelter Island, New York, following discussions of the problematic behaviour of cosmic-ray mesons, in which Marshak had developed a hypothesis that there were two kinds of meson, with a Yukawa meson decaying after a very short time into a lighter meson which was very inactive; and from the early observations Marshak and Bethe (1947) were able to deduce several properties of the π-mesons.

This possible relationship between two kinds of meson had also been discussed in Japan during the war (Sakata and Inoue, 1946).

A year later, π-mesons were produced by bombarding targets with particles from the Berkeley cyclotron, and it was confirmed that μ-mesons are only produced by the decay of π's. For the next few years cosmic-ray experiments contributed regularly to the list of known "elementary" particles, and though they made some of their greatest contributions, the complexity which they revealed called for more controlled experiments with artificial particle beams, as these became available. 1949 saw the simultaneous appearance of the neutral π-meson in experiments at the Berkeley accelerator and in cosmic rays (see Paper 10): in the atmosphere it plays an important part in starting electron–photon cascades, as it decays almost immediately into two γ-rays.

Enough of particle production was now known to connect the primary radiation with the radiation normally observed in the atmosphere. The resulting overall picture of the relations between the secondary components of cosmic radiation in the atmosphere will now be summarized.

3.7 Present picture of cosmic rays in the atmosphere

A proton entering the atmosphere suffers nuclear collisions with a mean free path close to 80 g cm^{-2}. However, the proton is not lost in the collision: it may emerge as a proton or a neutron, with reduced energy, and so the flux of nucleons of a given energy is found to fall off with atmospheric depth x as $\exp(-x/L)$, where $L = 120$ g cm^{-2} rather than 80 g cm^{-2}. Slow nucleons, particularly neutrons may be ejected from collisions.

If the incident kinetic energy $E \gtrsim 1$ GeV, π-mesons are likely to be produced, π^+, π° and π^- mesons appearing equally often on the whole, though when numbers are small conservation of charge somewhat favours the π^+ in proton collisions. Roughly $2E^{1/4}$ new charged particles are generated in a collision on average (E in GeV), though not all are pions. Because of their short lifetime ($2{\cdot}6 \times 10^{-8}$ s at rest) very few pions will be seen—only those produced in the last $\sim 3/4$ g cm^{-2}, for example, for 1 GeV pions at a depth 100 g cm^{-2}. The charged pions decay to muons: below 2·5 GeV the majority of muons generated near 100 g cm^{-2} will themselves decay or stop somewhere before reaching sea level; above this energy the time dilation factor is sufficient

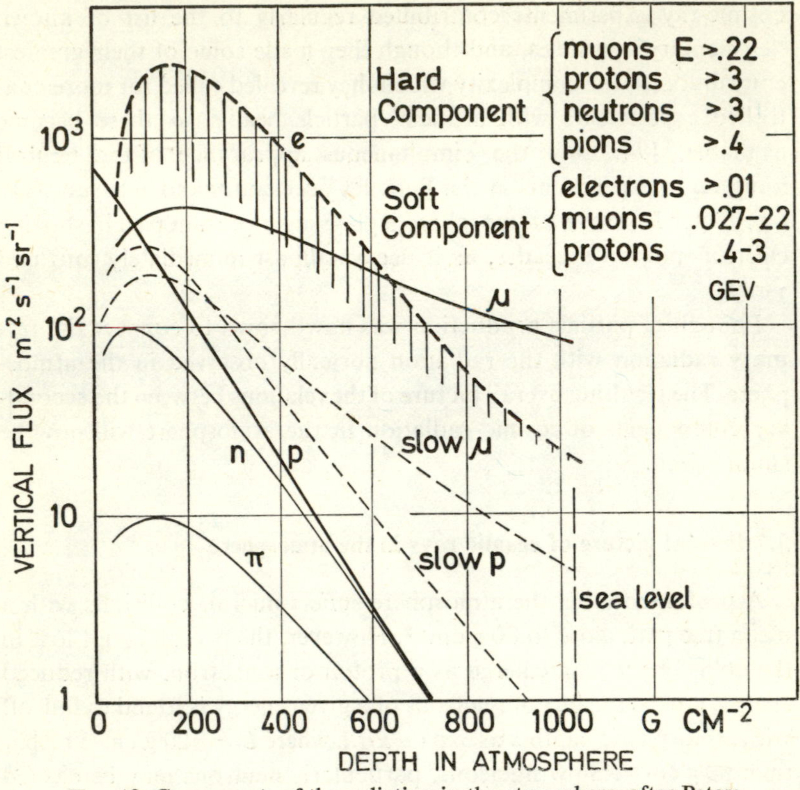

FIG. 12. Components of the radiation in the atmosphere, after Peters.

to enable the majority to survive so far. The resulting flux of muons, as it builds up and is then slowly attenuated, is shown in Figure 12. The decay process, e.g.

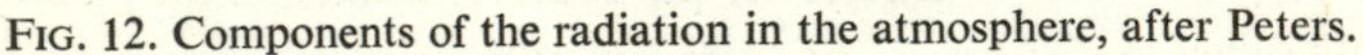

$$\mu^- \rightarrow e^- + \bar{\nu}_e + \nu_\mu,$$

generates electrons, which multiply by the cascade process and then die away. The flux of electrons in equilibrium with the flux of parent muons falls as the air density increases. In addition, the π° mesons decay:

$$\pi^\circ \rightarrow \gamma + \gamma$$

—and the photons generate electron–photon cascades which result in a

Table 3.2. The Discovery of the Elementary Particles

This table, an expansion of one given by Powell, Fowler and Perkins (1959), shows how and when the relatively stable elementary particles were discovered (antiparticles being included somewhat arbitrarily). The heavy lines show the discoveries made using cosmic rays. The particles are listed in order of increasing mass, except within charge multiplets.

Date: 1900, . . , 1930, 1931, 1932, 1933, 1934, 1935, 1936, 1937, 1938, 1939, 1940, 1941, 1942, 1943, 1944, 1945, 1946, 1947, 1948, 1949, 1950, 1951, 1952, 1953, 1954, 1955, 1956, 1957, 1958, 1959, 1960, 1961, 1962, 1963, 1964, 1965, 1966, 1967

Particle	Source of radiation	Instrument used	Specific observation made
$\bar{\nu}_e(\nu_e)$	nuclear reactor	liquid scintillator	Capture by proton
ν_μ	accelerator	spark chamber	Production of μ and not e
e^-	discharge tube	fluorescent screen	Ratio e/m
e^+	cosmic rays	cloud chamber	Charge, mass
μ^+, μ^-	cosmic rays	cloud chamber	Absence of radiation loss in Pb; decay at rest; mass
π^+	cosmic rays	nuclear emulsion	$\pi-\mu$ decay at rest
π^-	cosmic rays	nuclear emulsion	Nuclear interaction at rest
π^0	accelerator	counters	Decay into γ-rays
K^+	cosmic rays	nuclear emulsion	$K_{\pi 3}$ decay
K^-	cosmic rays	nuclear emulsion	Nuclear interaction at rest
K^0	cosmic rays	cloud chamber	Decay into $\pi^+\pi^-$ in flight
η	accelerator	bubble chamber	Total mass of decay products
p	discharge tube	spectroscopes; mass spectrometers	Charges and masses of ions
$\bar{p}$	accelerator	Cerenkov counter	e/m measured; annihilation
n	radioactivity	ionization chamber	Mass from elastic collisions
$\bar{n}$	accelerator	counters	Annihilation
Λ	cosmic rays	cloud chamber	Decay to $p\pi^-$ in flight
$\bar{\Lambda}$	accelerator	nuclear emulsion	Decay to $\bar{p}\pi^+$ in flight
Σ^+	cosmic rays	nuclear emulsion	Decay at rest
Σ^-	accelerator	diffusion chamber	Decay to $n\pi^-$ in flight
Σ^0	accelerator	bubble chamber	Decay to $\Lambda\gamma$ in flight
Ξ^-	cosmic rays	cloud chamber	Decay to $\Lambda\pi^-$ in flight
Ξ^0	accelerator	bubble chamber	Decay to $\Lambda\pi^0$ in flight
Ω^-	accelerator	bubble chamber	Decay to $\Xi^0\pi^-$ in flight
Very many "resonance" particles with lifetimes $\sim 10^{-23}$ to 10^{-19} s	accelerator	bubble chambers	Total mass of decay products
?"Fireballs"	cosmic rays	nuclear emulsion	Angles of meson emission
Quarks?	not found with accelerators; being sought in cosmic rays		Charge $\frac{1}{3}$ or $\frac{2}{3}e$

very large flux of electrons at high altitudes. The part of the electron flux arising in this way is shown shaded in Figure 12: only the unshaded part could be accounted for before neutral pions were discovered.

In the diagram full and dashed lines are used respectively for components contributing to the "hard" and "soft" radiation, divided roughly according to their ability to penetrate 10 cm of lead. The curves are derived from a summary by Peters (1958) with additions, and apply to a geomagnetic latitude of 45°, though they should apply also to higher latitudes, except perhaps at very high altitudes.

In the range of several GeV, nucleons and mesons have rather similar energy spectra and so their relative proportions vary rather slowly.

3.8 The use of cosmic rays to investigate high-energy interactions

Though the pion closed the chain of essential secondary radiations, work on the products of nuclear collisions intensified for several years, as several more unstable "elementary" particles quite unexpectedly came to light. In the year of the pion's discovery, the decay of the heavier K-meson was seen in cloud chamber photographs (Rochester and Butler, 1947)—an inverted V marked the decay of a neutral particle into two charged particles, without the mediation of any collision. Table 3.2 will suffice to indicate the contribution of cosmic ray work in this period. It was made clear that the simplest meson theory of nuclear forces must be greatly oversimplified, and that an immense wealth of detail awaited investigation in the field of elementary particles. After the commisioning of the "Cosmotron" accelerator in Brookhaven in 1952, which produced particle beams of a few GeV, and later the Bevatron at Berkeley, elementary-particle and cosmic-ray physics began to drift apart. The K-mesons and hyperons, which decayed with lifetimes $\sim 10^{-10}$ s were readily detected in cloud chambers, as they decayed after travelling a few cm from their point of generation; but only at the accelerators have controlled conditions yielded data on the many particles with lifetimes $\sim 10^{-22}$ s. Cosmic-ray conditions permitted measurement of masses, lifetimes and decay branching ratios, but did not favour determination of, say, magnetic moments, or very detailed interaction properties of particles. Only nuclear interactions at exceptionally high energies remained within the province of cosmic rays. Cloud chambers or emulsion chambers of many cubic feet are needed to collect the correspondingly low flux.

Paper 10 shows a characteristic appearance of such high-energy collisions. A narrow and a wider cone of emergent mesons, as seen there, are common features at such energies. The authors attribute the wider cone to further interactions of secondaries within the nucleus, which probably plays some part in the example illustrated, but a widely-held interpretation of many such events is that seen in the centre-of-mass system (c.m.s.), in which two colliding nucleons approach with equal and opposite momenta, the two nucleons afterwards recede with reduced energy, with "fireballs" trailing behind them—as though fragments of their surrounding meson clouds are pulled off by the collision—which then disintegrate isotropically into mesons, giving two moving centres of meson emission. The reader is referred to other reviews for details—e.g. Fujimoto and Hayakawa (1967) or Yash Pal (1967)—where he will find that different conclusions can be drawn: that several fireballs are formed at the highest energies, the fastest ones trailing just behind the nucleons—or that the mesons are mostly radiated from one fireball at rest in the c.m.s., but that the nucleons are excited to higher baryon mass states which then decay to a nucleon and 2 or 3 mesons, which together take most of the energy. Hence this field, in which progress is slow but may ultimately be important, will not be discussed.

Only a few parameters of the interactions may be deduced indirectly. For example, if at the top of the atmosphere the flux of nucleons of energy above E is $N(E, 0) = AE^{-\alpha}$, and in collisions, with mean free path λ, the nucleon emerges with a fraction β of its initial energy, it is easily shown that the flux of nucleons at a depth x g cm^{-2} in the air will be $N(E, x) = AE^{-\alpha} \exp(-x/L)$, where $L = \lambda/(1-\beta^{\alpha})$. Taking $\lambda = 80$, $L = 120$, $\alpha = 1{\cdot}63$ (Chapter 4), one finds that the elasticity $\beta = 0{\cdot}51$, a value which probably remains constant up to at least the highest energies studied in the emulsion experiments. Whether this reflects the fraction of the energy taken by a nucleon in baryon decay (in which case the relation for L requires a slight modification for the contribution of energetic pions to the flux of nuclear-interacting particles) or to a constant momentum transfer to the nucleon in all very high-energy impacts, remains to be tested.

Again, the observation that right up to the highest energies, there are about 25% more positive than negative muons also suggests that the proton shares its charge amongst a few favoured energetic pions, as in baryon decay.

But as yet the precise measurement of momenta of all particles involved in an interaction is far from the degree of perfection possible in accelerator conditions. We leave, therefore, the period which laid the foundations of high-energy and elementary-particle physics, to follow the study of the true primary cosmic radiation in the context of astrophysics rather than nuclear physics.

IV

The Primary Cosmic Radiation

As secondary rays are generated so readily it is necessary to go well above the altitude of the Pfotzer maximum to find the incoming cosmic radiation, well above 17 km (56,000 ft), that is, corresponding to a depth of 90 g cm^{-2}. Free balloons carrying apparatus which recorded either photographically or by radio telemetry readily reached over 90,000 ft (15 g cm^{-2}), and in the 1950s these became a matter of almost daily routine, level flights of several hours at up to 120,000 ft (5 g cm^{-2}) being quite possible. From 1950 rockets were available to take counters outside the atmosphere, and more recently, outside the magnetosphere.

As nuclear emulsions improved they served ideally as balloon-borne detectors, provided that the balloon's payload could be recovered after the flight, and that the background of tracks formed before the period of flight at high altitude could be minimized (for example by using interleaved emulsions which are automatically displaced just before and after the period of level flight). Their use has shown that the majority of the incoming primary particles are indeed protons, as foreseen from the work described earlier, but they at once led to the unexpected discovery that the incoming radiation includes quite heavy atomic nuclei, also travelling at relativistic speeds. Primary electrons or γ-rays are very few.

Three aspects of the primary radiation have been investigated: (a) the energy distribution of the particles, (b) the detailed composition —types of particle present and relative numbers—and (c) any departure from isotropy.

The interest in the latter in pointing to the direction of possible sources is obvious, but as yet the true primary radiation seems to be very nearly isotropic, after allowing for the effect of magnetic fields around the earth. It may at first seem surprising that so much effort has been expended over the years on detailed improvements in measurements of the flux, energy spectra (which after all turn out to show no line features) and composition, but as the radiation is so complex the details potentially contain much information: peculiarities of the composition may give clues to the place of origin and mode of production, and differences between the energy spectra of the components may call for explanation in terms of their history.

However, even the magnitude of the flux calls for some explanation, as the radiation carries at least as much energy as the light and heat from all the stars. A natural question arises: how typical is this of other localities in the universe?

4.1 Methods of determining the flux and energy spectrum

Many difficulties have hindered the accurate determination of the flux of primary particles. The geomagnetic method used by Clay to determine the energy spectrum has been widely followed, making measurements of the particle flux at different latitudes, as particles of magnetic rigidities from 15 GV downwards first penetrate the Earth's field. In his 1932 paper, Clay attempted a rough estimate of the energy spectrum from the radiation at sea level, but balloon observations near the top of the atmosphere are really necessary, for the effect each particle engenders at sea level is by no means independent of its energy and practically vanishes below 4 GeV. In fact this effect of atmospheric absorption was the reason given by Clay for the near-constancy of cosmic ray ionization at latitudes above $\sim 45°$ at sea level, as the lower energy rays allowed in at higher latitudes would only be detected at higher altitudes. In the summers of 1936 and 1937 Bowen, Millikan and Neher flew accurate ionization chambers to about 90,000 ft in India near the magnetic equator and at magnetic latitudes of 38°, 51° and 60°. The results are shown in Figure 13. Another flight at 56° and a similar one in 1937 by Carmichael and Dymond at 85°N showed that,

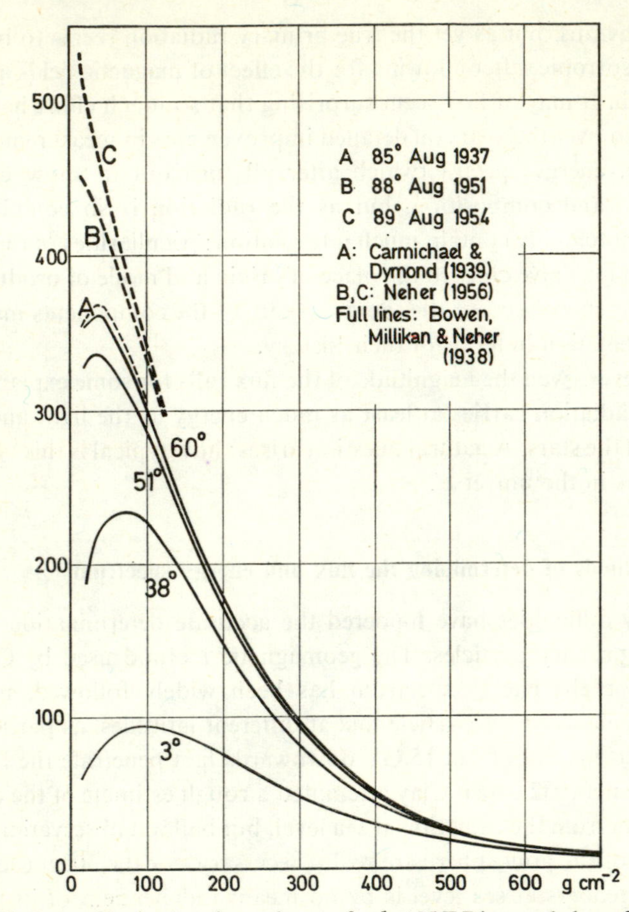

FIG. 13. Rate of ionization (ion pairs cm^{-3} s^{-1} at N.T.P.) recorded as a function of atmospheric depth in balloon flights at different magnetic latitudes.

even at the highest altitudes, there was no increase in flux beyond a latitude of 55°N, so that there were no primary particles with a rigidity below ~1·7 GV† (for protons, the corresponding kinetic energy ≃ 830 MeV).

The area under one of these graphs gives the rate of production of

† Allowing for non-dipole anomalies in the Earth's field.

ions in the atmosphere and hence the flux of energy into the atmosphere (taking 34 eV per ion pair); the area enclosed between two curves gives the energy flux in the incoming cosmic rays in the rigidity interval between what is allowed at one latitude and what is allowed at the next; hence an energy spectrum can be built up. This idea lay behind Millikan's balloon flights: he had from the start been inclined to believe that the spectrum should be discrete, and set out to define the energy bands. Figure 14, redrawn from his original results using revised cut-off rigidi-

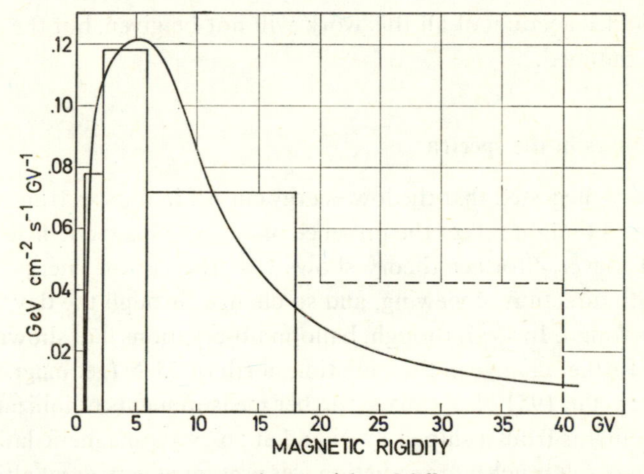

FIG. 14. Energy carried by incoming particles of different magnetic rigidity, after Millikan.

ties, shows the energy flux carried by cosmic rays of different rigidities, the rectangles corresponding to the areas between adjacent observational curves, and the smooth curve being drawn to have the same area. This method of approach is straightforward, but again not exact, for perhaps one-third of the energy (Puppi, 1956) may go into neutrinos and nuclear disruption, and not into ionization, and one has to take the cut-off rigidity for vertical rays as a rough effective figure to apply to the rays incident over all angles.

Telescopes of counters flown in balloons and rockets permitted direct counting of vertically-incident particles, so overcoming these two objections to the energy-deposition approach, but suffer from the difficulty that some secondary particles are detected even at the top of the atmosphere. This "albedo" radiation consists of particles shot back upwards ("splash albedo") and, in lower latitudes, an equal "re-entrant albedo" consisting of the splash from the same magnetic latitude in the opposite hemisphere, which has spiralled along a magnetic line of force.

At latitudes where all incoming particles must be relativistic, α-particles and heavier nuclei can be readily distinguished by the z^2 dependence of their ionization, and albedo gives no trouble.

A detailed account of all this work will not be given, but the results will be outlined.

4.2 Changes in the spectra

Janossy suggested that the low-energy cut-off in the spectrum shown in Figure 14 might reflect the presence of a dipole magnetic field of the Sun. However, Størmer theory shows that the cut-off energy would vary with direction of viewing, and so change through the day, which was not found. In 1937, though, balloon observations had shown there was no further change in the radiation north of 55°N (geomagnetic, of course); yet in 1951 Neher made further measurements of ionization in balloon flights from a ship at 88°N and at 56°N geomagnetic latitudes, and found that much more radiation was present at very great altitudes, giving a 25% increase in the total influx, though still with no difference between different latitudes beyond 55°. This was puzzling, but in 1953 Forbush discovered from long-term records that the intensity of cosmic rays at sea level showed a variation over the years related to that of sunspots. When Neher repeated the balloon flights in 1954 the primary flux was found to have increased considerably in latitudes north of 55° (shown by dashed curves *B*, *C* as compared with *A*, on Figure 13), and Neher reported (1956) that the experiments "support and extend the relationship found by Forbush, namely, that there is an inverse relationship between solar activity and cosmic-ray intensity. During the summer of 1954 the sun was at a minimum of activity. Ionization chambers sent up in the Arctic as well as at intermediate latitudes showed a considerable increase in intensity, during the period, compared with 1951. Atmospheric absorption, combined with magnetic rigidity requirements indicate that protons with energies down to at least 150 MeV were coming to the Earth in the summer of 1954. We conclude that there was no cutoff of primary particles, i.e. no 'knee', at least for protons down to this energy. Using the increased areas under the depth-ionization curves, combined with geomagnetic calculations, an estimate may be made of the numbers of primary particles coming in near the poles in 1937 (solar maximum), 1951 and 1954 (solar minimum). These values are 0·10, 0·14, and 0·24 cm^{-2} s^{-1} $sterad^{-1}$, respectively. The fluctuations in 1954 were also small compared with those measured in 1937, 1938 and 1951. The evidence points to low intensity and large fluctuations of cosmic rays when the sun is active and vice versa. These changes may be understood qualitatively by assuming a modulating mechanism in the form of (magnetized) clouds ejected from the sun and varying with the solar cycle." Evidently the lowest energy particles were most affected.

The Sun's magnetic and general activity varies with an approximately repeated 11-year cycle, with minima in the years 1913, 1923, 1933, 1944, 1954 and 1965. More low-energy particles manage to reach us in such years: whether all which are available do so is open to doubt.

4.3 The proton and helium components

Our basic quantitative information on cosmic radiation is contained in the energy spectra of the components. Though the protons form the larger part of the primary radiation, the precise measurement of their flux is hindered by the albedo. Thus although systematic observations with counter telescopes in balloons by Winckler *et al.* (1950) when extrapolated to the top of the atmosphere agreed well with Van Allen and Singer's (1950) measurements above the air in V2 rockets, these and other early measurements now seem to be high at some latitudes by a factor of more than 2. The albedo seems to be largely composed of relativistic electrons moving upwards, and perhaps returning in the other hemisphere, and since 1958 the problem has been largely resolved by the use of directional and velocity-sensitive detectors. In particular, an ingenious combination of a scintillator and a Čerenkov detector† introduced by McDonald and Webber (1959) distinguishes the protons from upward splash particles and α-particles, and has been used in a long series of balloon observations to determine the spectrum at different times during the solar cycle. Emulsions have also been used to measure the proton and the He flux: either these or scintillators can measure the ionization (or dE/dx) to give the energy of non-relativistic particles. In the kinetic energy range 0·5 to 15 GeV the geomagnetic latitude effect is used to measure the spectrum; continuation of the spectrum to higher energies depends on estimation of the energy released when individual protons make nuclear collisions in emulsions or "ionization calorimeters" or on working back from the spectra of the secondary mesons (Brooke, Wolfendale *et al.*, 1964). Since 1963 at the recent sunspot minimum period, very low energy particles have been detected in artificial satellites travelling outside the Earth's magnetic field: such particles can be stopped when their total energy and also their initial ionization (and hence charge) can be found by combining a thin and thick detector.

Figure 15 shows the proton spectrum in 1965, at the time of minimum solar activity, derived from several of the experiments mentioned. Above 350 MeV, the flux of protons with energy around E GeV is, to within the accuracy of observations, $J(E) = 18(E+2)^{-2\cdot 63}$ particles $m^{-2}\, s^{-1}\, sr^{-1}\, MeV^{-1}$, a representation proposed by Miyake, adjusted to refer to protons only.

The thinner line (P – 1959) in Figure 15 shows the form of the spectrum in 1959, when solar activity was high.

Primary He nuclei are easily distinguished by their heavy ionization from secondary particles and their flux has consequently been accurately measured for a longer period, using many types of detector, the accuracy being such that it revealed anomalies in geomagnetic thresholds. Results obtained near solar minimum are shown in Figure

† Čerenkov radiation: In a medium of refractive index n, a particle of charge ze and velocity βc emits light of intensity $\propto z^2[1 - 1/(\beta n)^2]$ if $\beta > 1/n$, at an angle $\sec^{-1}(\beta n)$ to its trajectory.

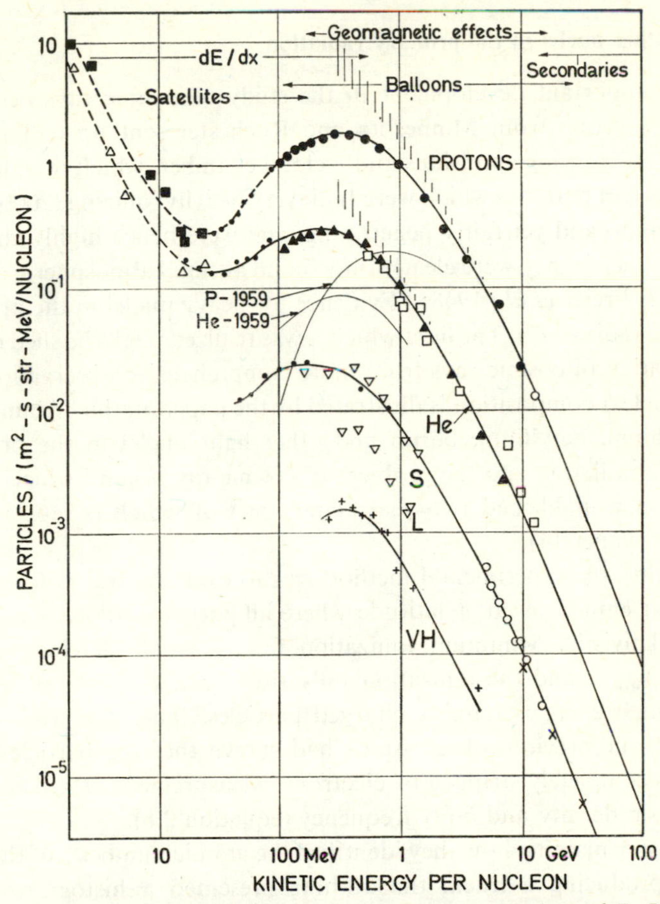

FIG. 15. Differential energy spectrum of primary protons and nuclei (L: Li, Be, B. S: $z > 5$. VH: $z > 19$) at a time of low solar activity. Methods of investigation are indicated for different energy regions. Very far from the Sun, the spectrum is probably as shown by the hatching.

15, classifying the particles by their kinetic energy per nucleon (i.e. $E/4$), again with a set of observations in 1959 (He – 1959). Differences in the changes in the proton and helium spectra have been the subject of much discussion, with regard to the mechanism causing the spectra to change.

4.4 Other nuclei in the primary radiation

An important development in the study of cosmic rays occurred when a group from Minnesota and Rochester sent up balloons to 90,000 ft carrying emulsions and a cloud chamber, which revealed the presence of particles which were both very heavily ionizing (2000 times minimum) and yet fairly penetrating, and were hence highly charged heavy nuclei: they were clearly only found at small atmospheric depths (Phyllis Freier *et al.*, 1948: "Evidence for heavy nuclei in the primary cosmic radiation"). The light which it was realized might be shed on the generation of cosmic rays from more comprehensive observations on the nuclear composition is illustrated by the paper on the "Abundance of lithium, beryllium, boron and other light nuclei in the primary cosmic radiation and the problem of cosmic-ray origin" which Bradt and Peters published two years later, part of which is reprinted in Part 2 (Paper 14).

In this, the experimental method was to examine tracks formed in nuclear emulsions at a latitude where all particles allowed in must be relativistic. Minimum ionization for a particle of charge ze is $z^2(I_{min})_{proton}$, and the ionization only rises by 13% to the plateau at very high energies. Highly charged particles, then, give very dense tracks, and previous observations had shown that the particles were atoms completely stripped of electrons. Measurement of z was based on grain density and δ-ray frequency (equation 3.8).

After explaining how they identified the atomic number z of the particle producing a track, the authors presented a histogram (their Figure 3) of the distribution in z, for 111 tracks having $z > 2$. Carbon and oxygen were prominent, but what will be termed light or L nuclei, with $z = 3$, 4 and 5 (i.e. Li, Be, B) were not numerous, and could mostly have arisen from the break-up of larger incident nuclei in collisions in the 20 g cm^{-2} of overlying air. Since an analysis of observations of nuclear collisions led the authors to accept that a nuclear collision of a heavier nucleus ($z > 5$) with hydrogen would on average yield at least 0·23 light nuclei of unchanged velocity (very recent accelerator experiments on the collisions of protons with nuclei would incidentally suggest $\sim 0{\cdot}34$ light nuclei per collision), it could be shown that after a very long passage through interstellar hydrogen cosmic rays must contain a much larger proportion of light nuclei than observed; hence they have travelled through rather little material, and their chemical composition still bears a strong relation to that of the radiation issuing from the source. It now seems that the detection of Li, Be and B in this experiment may have been a little inefficient (though there is little evident reason other than the additional requirement of uniformity demanded in accepting tracks of $z \leqslant 4$): the numbers of nuclei found are quite close to those generally quoted today at the top of the atmosphere; and rather more light nuclei are found at lower energies than at the 9 GV rigidity threshold appropriate to the latitude of their experiment. However, their main conclusion is little altered.

The upper limit deduced for the age of the particles, and other comments in the final section of this paper relate to ideas discussed in Chapters VI and VII, and their discussion is worth reading in connection with Fermi's paper; but the evidence that the radiation may contain not wholly blurred evidence of the chemical composition of the source region is obviously of great importance.

With the aim of separating, if possible, effects of the composition of the source, the preferential acceleration of ions of differing charge, and the time spent in interstellar space since acceleration, many more studies of the charge composition have been undertaken (the Minnesota groups still being very active in this field), improving the statistics, working close to the top of the atmosphere, and studying the detailed products of nuclear fragmentation which transform the charge distribution. In addition to emulsions, the Čerenkov–scintillator and other counters have been used with great success, being adaptable to radiotelemetry in balloons and in space probes beyond the magnetosphere. The charge resolution obtained by combining scintillator (its signal varying with z, β nearly as the ionization does) and Čerenkov signals is shown by Figure 16. The ellipses mark relativistic particles.

Energy spectra found in 1965 are shown in Figure 15 (L nuclei—$z = 3, 4, 5$; S nuclei—$z > 5$; VH nuclei—$z > 19$): apart from protons, all energy-per-nucleon spectra are remarkably similar (though a little

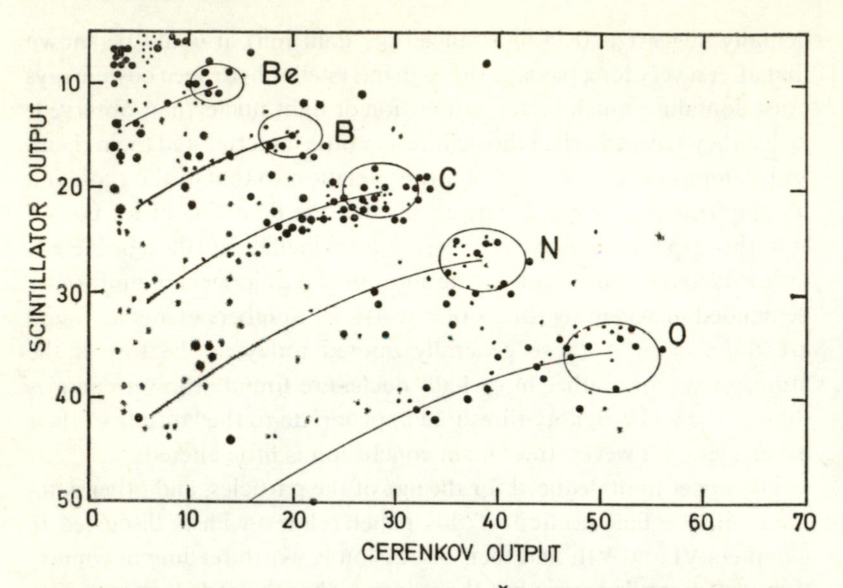

FIG. 16. Separation of nuclei by scintillator–Čerenkov counter combination (after McDonald and Webber, 1962). The smaller dots came from a flight where the threshold admitted slower particles, which extend further back away from the ellipses.

steeper for L, as remarked). During minimum solar activity, the fluxes of protons, α-particles and all other nuclei having rigidities 4·5 GV are 620±30, 94±4 and 8·7±0·3 $m^{-2} s^{-1} sr^{-1}$ respectively—in the ratio 6·6 : 1 : 0·093. Equal rigidity implies equal velocity for nuclei with $A = 2z$, and most comparisons are made on this basis: at high energies the ratio of proton to helium flux at equal velocity is 18 : 1. One may compare the ratio of abundance of H, He and all other nuclei in ordinary matter in the universe, 8 : 1 : 0·011 by number. Table 4.1 gives the relative numbers of 1000 nuclei with $z > 2$ (at equal rigidity) at 0·3 to 5 GeV per nucleon, derived from the more concordant of several recent experiments (Webber and Ormes, 1967; Mathieson *et al.*, 1968; O'Dell *et al.*, 1962; Anand *et al.*, Balasubrahmanyan *et al.*, 1965; and some of their later work), though the figures in several cases, especially for the adjacent very heavy nuclei, are much less well estab-

TABLE 4.1.
CHEMICAL COMPOSITION OF COSMIC RAYS AT TOP OF ATMOSPHERE

Element	Cosmic rays	Natural abundance†	C.R. source?
Li	50	} 0·0035 (Li, Be, B)	
Be	31		—
B	85		
C	285	148	310
N	75	104	
O	240	603	280
F	6	—	
Ne	46	69?	84
Na	13?	1	
Mg	50	24	84
Al	12	3	
Si	35	26	62
P	4	—	
S	10	10	30
Cl	2	—	
A	4	4	
K	—	—	
Ca	4	1	
Sc	1	—	
Ti	4	—	
V	2	—	
Cr	7	—	
Mn	3	—	
Fe	29	6?	100
Co	—	—	
Ni	2	1	

† Geometric mean of versions by Suess and Urey (1956), Cameron (1959).

lished than may appear from this table—three-figure accuracy (often two) is certainly not implied!

Li, Be and B are present, and assuming that they were virtually absent from the initial material, some fragmentation of nuclei has occurred, and one can refine the argument of Paper 14. Taking a collision cross-section πr^2, where $r = 1{\cdot}17(A^{1/3}+1)\times 10^{-13}$ cm for collisions of a nucleus (at. wt. A) with hydrogen, the effective collision mean free path for S nuclei (i.e. $z > 5$) in interstellar space is $\lambda = 6$ g cm^{-2}. As Bradt and Peters concluded that not all nuclei suffered collision, we suppose that there is another process removing them, with mean free path Λ. Then the proportion of initial S nuclei suffering collisions would be $\Lambda/(\Lambda+\lambda)$. For each S nucleus observed, there are 0·22 light nuclei, representing to a first approximation $0{\cdot}22/0{\cdot}34 = 0{\cdot}65$ collisions of S nuclei (neglecting collisions of collision products), in which perhaps 80% of the latter ceased to have $z > 5$, i.e. $\sim$0·52 S nuclei were lost. So the proportion of the original S nuclei which have interacted is $0{\cdot}65/1{\cdot}52 = 1/2{\cdot}3$, and thus $\Lambda \doteqdot \lambda/1{\cdot}3 = 4{\cdot}5$ g cm^{-2}. With a hydrogen density of 1 atom cm^{-3} as assumed in Paper 14, this would be traversed in 3 million years.

This lifetime is interestingly close to the half-life of the isotope Be^{10}, mentioned in Paper 14, but it seems that rather little of this isotope should be formed in fragmentation, and very painstaking measurements of the relative amounts of Be and B would be required (perhaps to 3%) to find the age of cosmic rays in this way.

The composition of cosmic rays does show a strong resemblance to the natural abundance of elements inferred in many parts of the universe, though with a notable enhancement of iron, and paucity of H and He, and possibly O. Attempts to allow for the effect of fragmentation *en route* (e.g. by Yiou *et al.*, 1968, shown in the table), though by no means unique, only strengthen these peculiarities of the presumed composition of the material at the source. Evidence that this is not due to a preference for larger charges shown by the accelerating mechanism is now accumulating from long flights of hundreds of square feet of emulsion by Fowler and co-workers, which show that nuclei with $z > 40$ do occur, but are thousands of times less abundant than iron, much as in natural material. A source region unusually rich in iron is thus favoured, as in a star at a late stage of its evolution—such as a supernova.

A pointer in the same direction is the suspected presence of unstable trans-uranic nuclei in the radiation detected by the large exposures referred to. (At the time of writing, this potentially exciting development cannot be properly evaluated, as a new method for detecting heavy tracks, using plastic sheets instead of photographic emulsions, indicates less extreme charges for the same tracks.)

4.5 Primary electrons and positrons

The complete absence of electrons from the primary radiation would be surprising, for mesons must be generated by collisions in the interstellar gas, decaying to electrons and positrons. Moreover, since the early 1950s, our understanding of radio astronomy has been based on the hypothesis that there are relativistic electrons in space. Not until 1961, however, when Earl flew a multi-plate cloud chamber in a balloon (at the geomagnetic latitude of Minneapolis, where there should be no trouble from re-entrant albedo at 1 GeV or so), and determined the energies of electrons from the showers they produced, was it possible to detect a small electron flux above the copious production of secondary electrons in the air. Meyer and Vogt also detected primary electrons with a counter system in the same year, and many other investigations have followed, using counters (measuring dE/dx and range, or shower size under lead) or spark counters, or emulsions. It has always been necessary to examine the behaviour of the flux as the top of the atmosphere is approached to check on the subtraction of the secondary electron flux, which is very considerable below 500 MeV, though above this energy it rapidly loses importance. In a few cases the east–west effect has been used to infer the relative number of electrons and positrons, and a magnet has also been used for this purpose: generally it seems that positrons are present but are outnumbered by electrons, which is taken to imply that much of the flux is of primary accelerated electrons, rather than products of meson decays from interactions in the interstellar gas.

Fluxes reported in the various experiments are shown in Figure 17, along with the positron–electron ratio. From 4 GeV upwards, the combined electron–positron flux is about 1% of the proton flux. Also shown is the predicted flux of secondary electrons and positrons from the interstellar interactions of protons; it tends to confirm the conclusion drawn from the charge ratio. (The flux of atmospheric secondaries at $\sim 3\ \mathrm{g\ cm^{-2}}$ depth also follows an almost identical curve to that predicted for the interstellar secondaries.) It is not yet clear whether the very low energy electrons detected in space probes originate relatively locally or not.

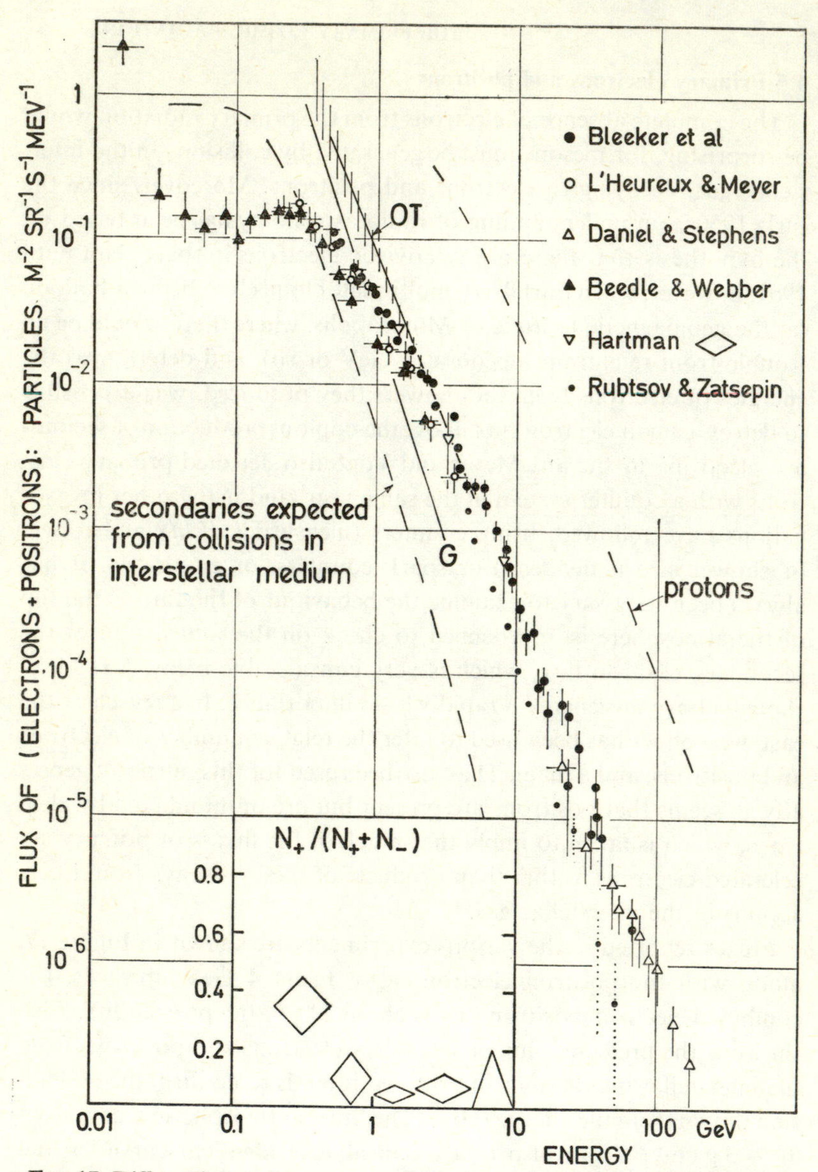

FIG. 17. Differential energy spectrum of the electron–positron component; and proportion of positrons as a function of energy. (Data largely from *Can. J. Phys.* 46 (1968), and Hartman (1967) and Anand *et al.* (1968).) Lines G and OT come from radio-astronomical data—section 5.6.

The results were obtained near the time of minimum solar activity, and the effect of the solar cycle is not yet known. If the theory of solar modulation described in Chapter VIII is correct and applies to electrons, the flux outside the solar system is probably within the limits indicated by the vertical shading (if the "modulation parameter" η lies in the range 0·6 to 1·0 GV).

The significance of the electron component will be discussed later.

V

Radio Waves from the Galaxy

5.1 The Galaxy

If we look beyond the solar system for the scene of operations of cosmic rays it is hard to find a smaller system than the Galaxy. Our galaxy is a characteristic spiral galaxy, perhaps type Sb, with a mass $1-2\times10^{11}$ times the solar mass, mostly in the form of stars, most of which are distributed in a disc of radius $\sim$15 kpc. (1 pc, or parsec, $=3{\cdot}3$ light years $\doteqdot 3{\cdot}1\times10^{18}$ cm.) For older stars the disc thickness is $\sim$700 pc (without a sharp edge), though it is thicker and denser near the centre of the system. Younger stars are found in a thinner disc and concentrated in flattened tubular "arms" which spiral outwards in the disc, as seen in photographs of extragalactic nebulae. Gas—mostly hydrogen—follows a pattern similar to the younger stars, with a mean density in our part of the disc of perhaps 2 atoms cm^{-3}, and a thickness between half-density points of $\sim$ 220 pc, though there are many clouds of greater concentration, and between arms the density is less. A few percent of the gas is fully ionized (H II), in regions near hot stars. The

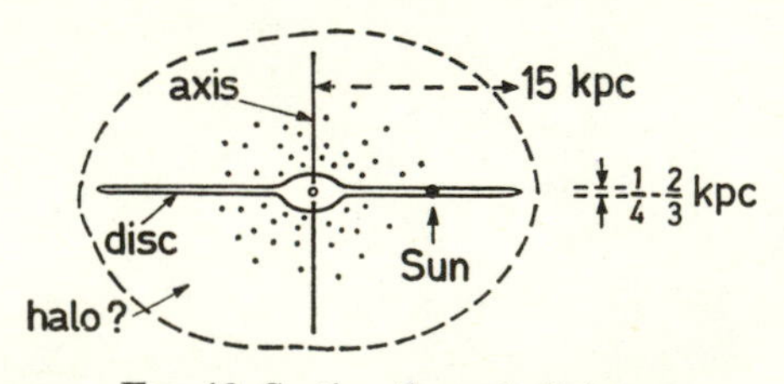

FIG. 18. Section through Galaxy.

Sun is about 10 kpc from the centre, and close to the central plane of the disc. There appears to be a second concentric but more nearly spherical system of star clusters and thinly scattered stars, apparently older (Population II), and a tenuous gas envelope probably also surrounds the visible galaxy. This large halo may be of importance to cosmic rays, as it greatly increases the volume of the system (to perhaps $\sim 3\times 10^{68}$ cm^3 very roughly), but its nature is still not well established. The galactic nucleus cannot be seen because of intervening dust clouds (see Figure 18).

5.2 Cosmic radiation—a solar, galactic or universal phenomenon?

Cosmic rays show no evident association with any local source; yet if they extend uniformly throughout all space their generation must represent the major process of energy production in the universe. The problem eases if the cosmic rays are largely confined to some smaller part of space. Explicitly, the energy density of cosmic rays away from the influence of solar modulation is probably $\sim$ 1·0 eV cm^{-3} (0·6 eV cm^{-3} actually reaches us when solar activity is a minimum). Compared with $2{\cdot}9\times 10^7$ eV cm^{-3} of ordinary solar radiation near the Earth, this is very small, but if the particles extend throughout the galactic disc a more suitable comparison is with the local density of starlight, $\sim$ 0·6 eV cm^{-3}—or, if universal, with the density of light from extragalactic sources, perhaps 0·006 eV cm^{-3}. The supposed primeval radiation (Chapter IX) is then the only rival, $\sim$0·4 eV cm^{-3}.

For this reason Alfvén surmised that there might be a weak magnetic field throughout the galaxy, probably of irregular form, as little as 10^{-9} gauss sufficing to hold most of the particles inside the disc for a very long time; at the same time their motions would be randomized towards isotropy. Even so, if the particles spend 1000 times longer in the disc than the thermal radiation (3×10^6 as against 2000 years), the cosmic rays represent a much greater fraction of the galaxy's radiation output than of the Sun's, which led Teller in 1948 to propose that cosmic rays should be considered as a much more local phenomenon. Richtmyer and Teller (1949) advocated a model in which the particles were trapped within 0·1 parsec of the Sun by a hypothetical magnetic field of 10^{-5} gauss, which was small enough to have escaped geomagnetic observation: in this way the storage gain factor might have been 10^7 rather than 10^3. Such a field could hardly retain protons above 10^{14} eV within the region, or iron nuclei above 3×10^{15} eV, but at that time this seemed adequate. Such models were quite attractive so long as cosmic rays could only be observed at the Earth, but fell out of favour in view of evidence to be presented in this chapter.

The interstellar and interplanetary magnetic fields indeed now appear to be as strong as this, but the latter seems rather to impede the approach of particles from outside. A short discussion of the nature of the postulated magnetic fields will give a useful orientation.

5.3 Cosmic magnetic fields

From 1948 onwards many authors explored the possibilities of magnetic fields for trapping and accelerating cosmic rays; later their importance for the slowing down of relativistic electrons was recognized. During this period interstellar space came to be seen as the site of more complex phenomena in which magnetic fields and turbulent gas motions play a significant part.

To explain this preoccupation with magnetic fields one must look on interstellar space as a good conductor of large dimensions. Then if one neglects the displacement current $\varepsilon_0 \dot{\boldsymbol{E}}$ in comparison with the conduction current, one has, using the conventional symbols,

$$\operatorname{curl} \boldsymbol{H} = \boldsymbol{j}, \qquad \operatorname{curl} \boldsymbol{E} = -\dot{\boldsymbol{B}}, \quad \operatorname{div} \boldsymbol{B} = 0,$$

and

$$\boldsymbol{j} = \sigma(\boldsymbol{E} + \boldsymbol{v} \times \boldsymbol{B}),$$

where $\boldsymbol{j}$ is the current density in a medium of conductivity σ moving with velocity $\boldsymbol{v}$.

Hence

$$\dot{\boldsymbol{B}} = -\operatorname{curl} \boldsymbol{E} = \operatorname{curl}(\boldsymbol{v} \times \boldsymbol{B} - \boldsymbol{j}/\sigma). \tag{5.1}$$

Consider first the meaning of this if the conductivity is so large that the last term may be neglected, and consider the magnetic flux through a closed loop which is carried along by the medium. The flux Φ will change because of changes in $\boldsymbol{B}$ and because of movements in the boundary of the loop. Hence, integrating over the loop,

$$\frac{d\Phi}{dt} = \int_S \dot{\boldsymbol{B}}.\boldsymbol{dS} + \int_l \boldsymbol{B}.\boldsymbol{v}\times\boldsymbol{dl},$$

where $(\boldsymbol{v}\times\boldsymbol{dl})\,dt$ is the area added to the loop by the motion of the element $\boldsymbol{dl}$ of the perimeter.

$$\frac{d\Phi}{dt} = \int_S \dot{\boldsymbol{B}}.\boldsymbol{dS} - \int_l \boldsymbol{v}\times\boldsymbol{B}.\boldsymbol{dl} = \int_S [\dot{\boldsymbol{B}} - \text{curl}\,(\boldsymbol{v}\times\boldsymbol{B})].\boldsymbol{dS},$$

$$= 0 \quad \text{if} \quad \boldsymbol{j}/\sigma \ \text{ is negligible.}$$

In this case, the magnetic lines of force move as though they are attached to the material: the flux through any portion of matter is constant. Matter may easily move along the lines of force, but if ionized matter moves in any other way it drags the lines with it. The flux is "frozen in", in Alfvén's description. A consequence of this is that turbulent motion of interstellar matter will stretch and entangle lines of force, thus amplifying any small initial magnetic field, and increasing the magnetic energy. Ultimately the magnetic forces will retard the motion: an approximate equality of magnetic and kinetic energies may be attained, and although it is hard to derive an exact rule, such an equipartition in interstellar space has often been assumed in order to anticipate the strength of magnetic field present. Judging from observed velocities of turbulent motion one might expect about 5 μG on this basis.

The complete freezing-in of flux, however, depended on the resistivity being negligible. If we return to equation (5.1), considering now a coordinate frame in which $\boldsymbol{v}$ disappears, we have, if for simplicity we consider σ to be constant:

$$\dot{\boldsymbol{B}} = -\frac{1}{\sigma}\,\text{curl}\,\boldsymbol{j} = -\frac{1}{\sigma}\,\text{curl curl}\,\boldsymbol{H} = \frac{1}{\sigma}\,\nabla^2\boldsymbol{H},$$

since grad div $\boldsymbol{H} = 0$.

$$\dot{\boldsymbol{B}} = \frac{1}{\mu_0\sigma}\,\nabla^2\boldsymbol{B}.$$

This equation describes a diffusion of the magnetic field, and if the characteristic distance over which the field changes is l, $\dot{B} \sim B/(\mu_0 \sigma l^2)$, giving a mean lifetime $\tau \sim \mu_0 \sigma l^2$ for dissipation of the field. σ may be typically $\sim 10^3$ (ohm-m)$^{-1}$ in an interstellar plasma—not unlike a metal—so for a cloud 1 light year across one might expect $\tau \sim 3 \times 10^{21}$ years. Really the situation is more complex, as the magnetic field hinders the motion of ions across the lines of force; in cold, little ionized, interstellar gas clouds the dissipation time is considerably reduced, perhaps to $\sim 3 \times 10^8$ years (Cowling, 1957). Nevertheless, the magnetic flux changes so slowly that it is usually regarded as frozen-in.

Another consequence of this magnetic inertia is that electrostatic fields are of little direct importance, as on any ordinary time scale the initial current flowing in response to an applied potential will easily induce a nearly equal back-e.m.f., an effect compared by Alfvén to the skin effect at high frequencies: the inductance determines the current in a circuit for a short period (i.e. $< 10^{10}$ years in some cases), and electric fields cannot be derived from a potential.

In some cases plasma instability and turbulence may reduce the field inertia.

5.4 Cosmic radio waves

In 1932 Jansky unexpectedly detected radio waves coming from the Milky Way, even though he could not detect radiation from the Sun, using a receiver tuned to 20 MHz. Surprisingly, this discovery was not followed up until an amateur (Reber, 1940) managed to detect similar radiation at 160 MHz, and began to map out the source distribution in the sky. Directional resolution was better at short wavelengths, and immediately after the world war in 1945 the improved short-wave techniques were applied to follow up these observations, by Hey and others, and large aerial arrays were also built for work at lower frequencies. "Radio astronomy" sprang to life, and within a few years the general pattern of radiation over a frequency range from 10 MHz to 3000 MHz was known. The most prominent feature was an oval concentration in the direction of centre of the Galaxy, with a band a few degrees wide around the remainder of the galactic plane, much less intense away from the centre, tailing into a lower background of radiation. Further details emerged later, but the most surprising feature was the very high intensity of the radiation, particularly at the lowest frequencies, where it corresponded to a black-body brightness of $T_b = 2\times10^5$°K in the galactic plane—whereas the hottest extended sources visible optically were localized H II clouds at $1{\cdot}2\times10^4$ degrees.

The spectrum of radiation from a hot gas, the simplest type of source, can be described by the energy flux I_ν per unit frequency interval. For radio frequencies, where $h\nu \ll kT$,

$$I_\nu = \frac{2\nu^2}{c^2} kT_b$$

if the emitter is a complete absorber, at temperature T_b: hence the effective temperature T_b is derived. But by 1950 observations covering the frequency range 20 MHz to 480 MHz showed that I_ν, far from increasing as ν^2, decreased somewhat with increasing ν. For a transparent hot gas, of optical thickness τ, the effective temperature is $T_b = T(1-e^{-\tau})$, or $T\tau$ if $\tau \ll 1$, and for hot interstellar gas, $\tau = 0{\cdot}010T^{-3/2}\,[17{\cdot}7+\ln(T^{3/2}/\nu)]\left(\int N_e^2\,dl\right)\nu^{-2}$ (Ginzburg and Syrovatskii, 1964), where N_e is the number of free electrons per cm³, and the integral is carried out along the line of sight (dl in cm). Hence I_ν would be independent of ν at high frequencies, though H II clouds close to the galactic plane become optically thick for ν below a few MHz.

In a region within 100 pc of the galactic plane, thermal emission and absorption can be seen at very high frequencies, but not in general; and particularly at 20 MHz, the intensity implies a mean kinetic temperature very much above 10^5 degrees, and the visible hot ionized clouds are not only much cooler, but localized.

There was thus a problem in accounting for the galactic radiation. Hot diffuse gas having failed, it was widely supposed that the radio galaxy was made up of numerous "radio stars" in analogy with the visible star clouds of the "Milky Way". But, if so, the nearer individual "radio stars" should have stood out. Some intense "point" sources were indeed seen, and at first a misleading flickering suggested stellar outbursts, but a few were soon identified as very different objects—notably distant extragalactic nebulae and the unique Crab nebula.

In a short note reprinted in Part 2, Alfvén and Herlofson (1950: Paper 13) proposed another radiation mechanism: electrons circling in magnetic fields would radiate unusually strongly if they were highly relativistic, and at a few hundred MeV could radiate in the appropriate frequency range. They had in mind the Teller model of cosmic rays trapped in a region around the Sun, extending the idea to stars, some of which might have stronger trapping fields; thus a model of a "radio star" was proposed, assuming that at energies below 1 GeV some electrons were present in the cosmic rays. On seeing this, Kiepenheuer (1950) showed that numerous "radio stars" were unnecessary: this radiation process could account roughly for the intensity of galactic radio emission if one simply assumed that the cosmic radiation extended throughout the Galaxy and $\sim 1\%$ of the particles were electrons, which moved in interstellar magnetic fields of $\sim\mu$gauss. (There are, nevertheless, some objects to which Alfvén and Herlofson's model applies—e.g. the Crab nebula.) The radiation process is referred to as synchrotron radiation by Western physicists, after a theoretical treatment developed by Schwinger (1949) in another context: Russian physicists have used the term magnetic bremsstrahlung (or magneto-bremsstrahlung). At the time, no electrons had been detected in the primary cosmic radiation.

This suggestion was taken up and developed by Ginzburg and Shklovskii, who showed that radio-astronomy could give quantitative information about cosmic rays in regions distant from the Earth, thus vastly improving our chance of understanding their origin. Ginzburg showed that the radio-frequency spectrum was simply related to the energy spectrum of the radiating electrons. An energy flux spectrum $I_\nu \, d\nu \propto \nu^{-\alpha} \, d\nu$ is generated by electrons with an energy spectrum $J(E) \, dE \propto E^{-\gamma} \, dE$, where $\gamma = 2\alpha + 1$. Values deduced for γ have fluctuated a little: of late 2·2 to 2·4 has been favoured, which is very reasonable for a cosmic-ray spectral index (near 1 GeV).

The question of the distribution of cosmic rays in the Galaxy will be taken up later: more important was the observation of a very high concentration of relativistic electrons in the Crab nebula, and in related objects, which gave rough quantitative support to the view that supernovae were effective sources of cosmic rays.

These deductions are outlined in a paper by Ginzburg (1956) entitled "The nature of cosmic radio emission and the origin of cosmic rays", reprinted as Paper 15, which summarizes the results of his earlier work.

5.5 The Crab nebula

Taurus A, one of the few intense radio sources of small diameter identified with a visible object within the Galaxy, coincides with a most remarkable gaseous nebula—the Crab nebula. Many fine pincer-like filaments of excited gas are seen superimposed on a woolly background which emits a continuous spectrum of light, the whole system expanding at high speed from the site of a stellar explosion, the supernova observed in 1054. It is believed to be about 2 kpc away.

Again, an impossibly high temperature would be needed to account for the radio noise (or the continuous light) as a thermal emission, and Shklovskii suggested that both radio noise and visible light were produced by the magnetic bremsstrahlung process, as it appeared possible to draw a smooth curve through both parts of the spectrum (see Figure 19). This rare instance of optical radiation provided a test case for the proposed mechanism, for Gordon (1954) and Ginzburg (1954) pointed out that the light should then be noticeably polarized, and observations by Dombrovsky showed this to be the case. (The electric vector of the radiation is aligned parallel to the projected direction of the electron's acceleration at the time of emission, which is normal to the magnetic lines of force, as shown in Figure 20. Strong radiation is only emitted nearly tangentially to the particle's motion; for at this point there appears to be a sudden jerk in the motion, when one allows for the time taken for propagation of light or electromagnetic potentials to the observer.)

Up to this time, only the Russian workers had taken the synchrotron radiation mechanism seriously, but Oort was then shown the work, and in Leiden he began detailed observations on the polarization.

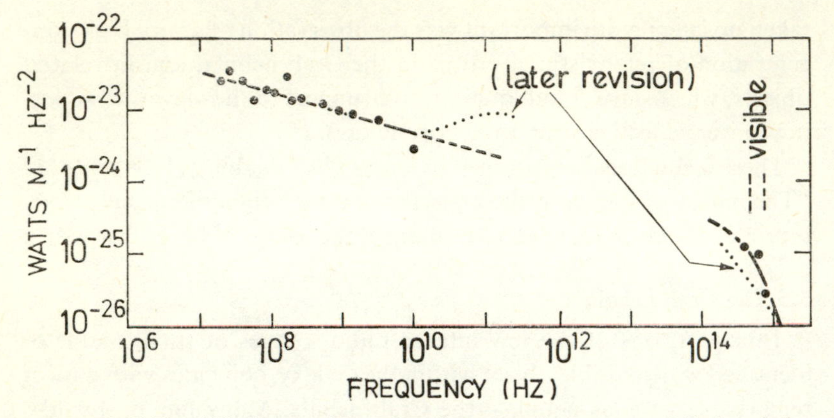

FIG. 19. Shklovskii's juxtaposition of radio and optical spectra of Crab nebula.

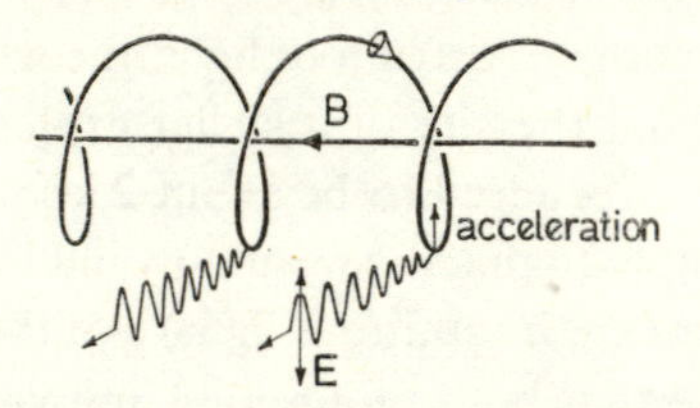

FIG. 20. Polarization of radiation from relativistic electron in magnetic field.

Later Baade (1956) took photographs through polarizing filters with the 200 in. telescope, and the light was clearly shown to be highly polarized with patterns revealing loops of magnetic field, and giving very good backing to the theory of emission by synchrotron radiation.

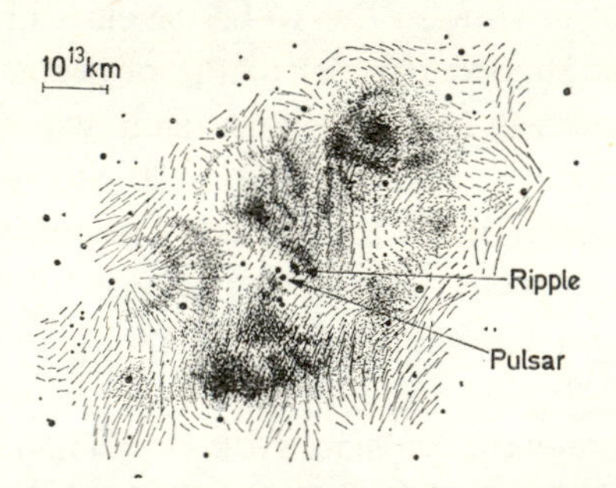

FIG. 21. Patterns of polarization of light of Crab nebula, after Baade and Woltjer.

(Figure 21 is a drawing of these polarization patterns, based on Woltjer's analysis of these photographs.) Faraday depolarization and poor resolution hinder a similar experiment at radio frequencies.

In the reprinted paper, Ginzburg estimates the energy content of the relativistic electrons in the Crab. The result hangs on the assumed strength B of the magnetic field: in general the radio intensity is proportional to the product of the electron flux and $B^{(\gamma+1)/2}$, assuming γ is known from the spectrum ($\simeq 1{\cdot}6$ in the case of the Crab). It has usually been argued that the pressure of the magnetic field must approximately equal that of the "cosmic-ray gas" or, approximately, that their energy densities are equal, if the cosmic rays do not break the system open and are accelerated by the turbulent field, and this supposed equality has been used to estimate B in several sources (though the magnetic pressure might be higher). Ginzburg assumed a rather conservative value of 10^{-4} gauss in 1956: more recently, the point at which the radiation spectrum steepens has been connected with energy losses of the electrons and hence with the field strength, suggesting $\sim 6\times10^{-4}$ gauss for the Crab.

Unfortunately this remarkable method of deducing the cosmic-ray electron flux in such an object gives no hint of the flux of protons and nuclei. One may assume that electrons form 1% of the total flux, as near the Earth, and on this basis the Crab nebula would contain 10^{49}–10^{51} ergs in the form of cosmic rays according to Ginzburg's figures; but the effect of the pressure of so high a flux would have been very noticeable, and in the Crab nebula at any rate the supposed proton flux cannot greatly exceed the electron flux ($\sim 4\times10^{47}$ ergs).

Several other supernova remnants have since been recognized as radio emitters, and a typical cosmic ray content, estimated similarly (with the 1% assumption), is a few times 10^{49} ergs.

A noteworthy point is that the energy of electrons giving the optical emission is, from Ginzburg's equation (4), $\sim \sqrt{[\nu/(1{\cdot}4\times10^{6}B_{\perp})]}\times 0{\cdot}51$ MeV, i.e. (3 to 5)$\times10^{11}$ eV—well into the cosmic-ray energy range. Apparently the energy spectrum steepens above 10^{11} eV, and although interstellar absorption of the light exaggerates the apparent steepening, the effect has been attributed to energy loss due to radiation, which in a period ~ 100 y would be large for electrons in this energy range (e.g. from equation 1 of Alfvén and Herlofson).

It is perhaps unfortunate that so much weight has been placed on this unique object, which may not be typical of supernovae. The energy spectrum index is unusually small, and the nebula is an unusual X-ray source. However, it has served to establish the significance of synchrotron radiation, and to demonstrate that there are local dense concentrations of cosmic rays, surely sources; whilst the recent discovery that the central star long suspected to be the supernova emits pulses both of radio and optical emission (pulsar NP 0532) promises further revelations.

5.6 Radio emission from normal galaxies

Supposing that relativistic electrons with an energy spectrum $N(E)\,dE = KE^{-\gamma}\,dE$ fill the galactic disc, and that magnetic fields are present, Ginzburg calculated the radiation intensity to be expected (equation 7 of his paper): the rough energy spectrum which he deduced from the early radio intensity measurements may be compared in Figure 17 on p. 68 (line G) with the electron flux later actually observed near the Earth. Radio-astronomical techniques have developed greatly since then, and the agreement is improved if we apply Ginzburg's equation (7) to recent data. Thus according to Okuda and Tanaka (1968), at 404 MHz the radiation emitted per kpc length of space in the galactic disc near the Sun corresponds to a brightness temperature $T_b = 3°$, and the intensity varies as $\nu^{-0{\cdot}7}$. If we assume a chaotic field of strength $B = 4{\cdot}5 \times 10^{-6}$ gauss (effective $B_\perp = 3{\cdot}6\ \mu$G), K in the electron spectrum may be deduced, and converting from electron density to flux (and ergs to GeV) the result is the line OT in Figure 17. It is higher than the observed electron flux, but very close to that extrapolated (hatched area) beyond the domain of solar modulation.

We may accept then, that the radio-astronomical observations can map out the product [cosmic-ray electron density$\times B^{1{\cdot}7}$], taking $\gamma = 2{\cdot}4$. This product seems for example to be 10 times greater at the ecntre of the Galaxy than in our neighbourhood.

Shklovskii drew attention to the non-isotropic distribution of the background radiation away from the galactic plane: even here it is stronger towards the galactic centre; he interpreted this as radiation from a large spherical corona or halo centred on the Galaxy and evidently containing a finite electron density and magnetic field. It is unfortunately hard to see this region clearly, as much of the radiation is probably from local wisps emerging from the disc, but an effective radius of about 10–15 kpc is probable. Observation of a similar halo around the great Andromeda nebula M31 by Baldwin (1954) gives a clearer picture of the halo density falling steadily away from the plane of that galaxy, dropping to half at 10 kpc. Cosmic-ray particles may thus extend well out into a near-spherical volume.

Though they tell us little about protons and nuclei, radio-telescopes can map out the electron component at even greater distances. Ginzburg estimated the energy content of our galaxy in cosmic rays as 2×10^{56} ergs, and it is notable that many other galaxies within a few Mpc show very similar total energies, estimated from their radio emission, and in their central regions most have spectral indices γ close to 2·2, if no correction for radio absorption is required. It is worth remarking that in some theories of cosmic rays, the value of γ would be almost accidental. With very large aerial arrays resolution is such as to show that the emission from many galaxies exhibits a central core $\sim$ 1 kpc across, which is surrounded by a more extended region in which the spectrum is steeper, as though electrons diffuse out from the core, and the more energetic ones are depleted by radiating away their energy; but the relative prominence of the core and disc is variable (e.g. Kuril'chik, 1965). It is not clear whether halos are generally present. We have thus come far from a local model, and these general features call for an explanation.

5.7 Unusual radio sources

Although they are not mentioned in the selected papers, certain very remarkable types of radio source have attracted attention. Firstly, the radio galaxies. The second strongest "point" source of radio emission was identified as an unusual galaxy at the great distance of 220 Mpc (within which distance there may be $\sim 5 \times 10^6$ galaxies), and many other very powerful radio galaxies are known. At first these were thought to be galaxies in collision, but this interpretation was later abandoned. A common feature is that the radio emission comes from two, or more, large zones on either side of the plane of a visible galaxy, as though two energetic plasma clouds had been ejected from the galactic centre, probably along a magnetic axis, though the dynamical difficulties involved in this have been pointed out by Burbidge. If this interpretation is right, the energy of the relativistic particles is 10^4 times that of a normal galaxy. At only 3 Mpc away, the galaxy M82 shows what may be an early stage of such a system, for direct photographs show that $\sim 0{\cdot}05\%$ of its mass projects out from the disc as filaments of hydrogen, travelling at up to 10^3 km s^{-1}. The light in these filaments is polarized, evidently synchrotron radiation from 10^{13} eV electrons, and shows that the filaments are running along lines of force. Evidently their origin was an explosion in the galactic core, about 2×10^6 years old.

Since 1960 the sources known as quasars have been studied, but to little avail as yet. Their optical spectra are most unusual and have not been satisfactorily explained, hence their energy content is subject to revision but is probably similar to that of the powerful radio galaxies; however, they are probably much more thinly scattered in space.

Finally, on quite a different scale, was the discovery by Hewish *et al.* (1968) of pulsars, the radio sources which emit short bursts of radio emission at very precisely defined intervals, ranging from 33 ms to 4 s, and by contrast only a few hundred parsecs away, and clearly related to the plane of our galaxy. In an account of the established achievements of cosmic-ray physics, speculation on such recent work is out of place, but, as will be seen, they could be very relevant, for the generation of the electromagnetic pulses evidently calls for strong particle fluxes and fields.

VI

Extensive Air Showers

6.1 Particles of very high energy

The discovery of air showers in 1938 implied that the energies of cosmic-ray particles extended several orders of magnitude above those deduced from the geomagnetic effects.

It arose out of experiments on the nature of particle interactions in which small showers of particles were ejected from metal plates, as discussed in Chapter III. Auger and Ehrenfest (1937), for instance, from cloud chamber photographs taken at the Jungfraujoch observatory, concluded that the showers were well described by the recent cascade theory of Bhabha and Heitler, and that the "soft component" of cosmic rays consists of shower producing photons and electrons which are themselves shower particles produced higher in the atmosphere. Showers were sometimes seen entering the cloud chamber, but their scale was not then suspected.

Another method of investigating shower generation was introduced by Rossi, who modified the counter-coincidence method of Bothe and Kolhörster by adding a third counter placed so that at least two particles had to arrive to discharge all three counters. Although the object was to study the production of secondary particles in various materials, it was often noticed that a few coincidences could still be observed with nothing placed above the counters—evidently showers from the air, which could be seen with counters up to 40 cm apart (Bothe *et al.*, 1937). Maze, working in Auger's group in Paris, improved the method of recording the simultaneity of counter discharges by shortening the voltage pulses to give a resolving time of a few microseconds, as compared with milliseconds in the earliest experiments, thus reducing the rate of "accidental coincidences" caused by two unrelated particles to less than 1 per hour. Truly related particles were then found even with counters several metres apart (Auger, Maze and Grivet–Meyer, 1938), and in the summer of 1938 the group energetically extended their observations by coupling up counters in light-roofed huts much further apart, finding coincidences exceeding the accidental rate at up to 300 m, and making experiments at two mountain stations. A report on this work, given at a conference in Chicago in the following year, is reprinted as Paper 7.

Kolhörster, Matthes and Weber (1938) working along similar lines reported detecting counter coincidences at distances up to 75 m.

Now there were evidently far more particles present than just the two which were detected, for the known total flux of particles could not contain so many independent pairs: there was a large shower of particles—an "extensive air shower". As Auger indicates, the density ϱ with which particles arrived on the ground in one of these events could be roughly determined by finding the effect of adding a third or fourth counter: the likelihood of an additional counter of area A being struck is $(1-\exp(-A\varrho))$. It will be seen from Paper 7 that although the true extent of the showers was not well defined, from the observed density and extension there must have been $\sim 10^6$ particles in the large showers.

In Manchester, Janossy and Lovell (1938) also photographed dense showers in a cloud chamber with a lead plate, and concluded like Auger that the tracks could be interpreted as electrons and positrons.

Thus the observations seemed best described by the theory of multiplicative electron–photon cascades, with an electron (or positron or photon) of very high energy starting a cascade at the top of the atmosphere, which then built up to a maximum number of particles part way down, and would be dying away in the lower atmosphere. This view of the nature of the initiation of the shower was mistaken, but this does not invalidate Auger's remarkable conclusion about the energy: simply taking the average particle energy to be about the critical energy

(87 MeV for air), and considering the energy expended in ionization in the atmosphere, he estimated the total energy of the shower, and hence of the primary particle, to be $\sim 10^{15}$ eV.

This discovery extended the energy scale of known radiations by an even greater factor than did the original discovery of cosmic rays, posing new problems with regard to their origin. Auger also notes that the energy distribution of these primary particles appears to be a natural continuation of the previously known energy spectrum, suggesting that the particles are probably of the same nature as the ordinary cosmic rays, and at the time this obscured the evidence that ordinary cosmic rays were protons (although Hilberry (1941) argued that it meant shower primaries were not as thought).

Actually, in these first experiments, Auger *et al.* noticed several features of showers which eventually pointed to a different origin. For one thing the presence of a few percent of penetrating particles is suspected; and later Cocconi, Loverdo and Tongiorgi (1946) and Daudin (1945), reporting experiments in the Alps during the war, established that a few percent of the particles are muons, which are spread more widely than the electrons. The Italians were able to conclude that many mesons must be formed in the early high-energy interactions, which would not be the case in ordinary electromagnetic cascades. A few protons and neutrons were also found. Skobelzyn *et al.* (1947) also found that coincidences could be obtained with counters up to 1 km apart, which they could not reconcile with the expected spread of electron–photon showers.

It gradually became apparent that these electron–photon cascades were generated by another component; some regenerating agent was preventing the showers from being attenuated as rapidly as expected from shower theory (and the attenuation rate quoted by Auger may be compared with the ~ 120 g cm^{-2} attenuation length for energetic protons). More recently it has been shown that a single nucleon of energy $\sim 10^{12}$ eV is often present at the centre (e.g. Dovzenko *et al.*, 1960). The first demonstration of how a (small) air shower starts was given in the reprinted paper (no. 10) by Kaplon, Peters and Bradt on "Evidence for multiple meson and γ-ray production in cosmic-ray stars", with a photograph of a shower being started by an α-particle. Hence air showers are regarded as the products of very energetic protons or nuclei.

The importance of the discovery by Auger, Maze and their colleagues, is that it makes possible the detection of such high-energy particles tens or hundreds of metres from their actual trajectory, so that their very small flux becomes measurable. Many air shower detectors have been set up since then, with the aim of detecting the arrival of the primary particles, so as to determine the extent of the cosmic ray energy spectrum and to find the directions of arrival of particles too energetic to have been readily deflected on the way from their sources. Another objective has been to unravel the cascade to find out about particle interactions at ultra-high energies, but this has not made much headway yet.

Because the showers extend over such a large area, and arrive randomly, it is only relatively recently that the detailed form of a shower has been clearly demonstrated. If only a few particles are detected, it is not clear whether they are close to the centre or not, but since 1953 several Russian groups, particularly under Vernov, Khristiansen, Nikolskii and others, have developed detectors in which several hundreds, and later thousands, of counters were hodoscoped to show particle densities in detail within a few metres of the axis of small showers, and Fukui, Miyake, Suga and others in Japan have used neon flash tubes and discharge chambers to photograph the particle distribution over an area of a few square metres.

Auger had remarked on local concentrations of particles: in fact there is generally seen to be a single region of high density at the centre from which the density falls away simply as a function of distance (Figure 22). Rather surprisingly, the distribution is hardly distinguishable from the form calculated for a simple electron–photon cascade, where the spread is due to multiple coulomb scattering. The original π-mesons which decay into γ-rays are apparently emitted with small transverse momenta, so that they do not aim far enough from the axis to make distinct multi-cored showers, and most showers of $\sim 10^5$ particles appear single-cored to well within 1 m of the axis, and so the

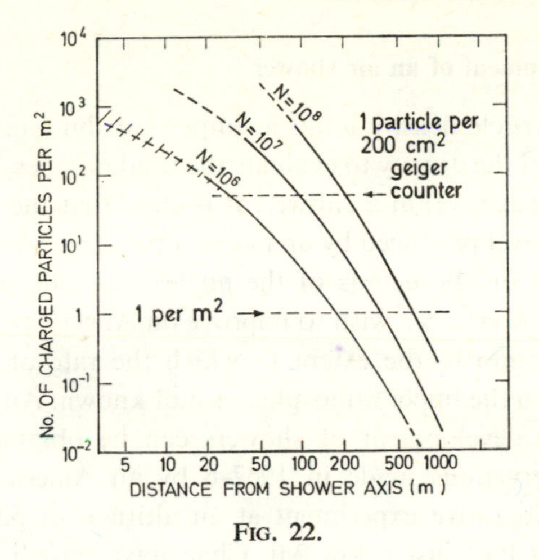

FIG. 22.

lateral distribution of the electrons gives little information about the angles of production of mesons.

However, some groups find that within ~ 2 m of the axis, the particle distribution varies from shower to shower, as though it depends on the amount of energy still present in the nucleon a few km above the apparatus; and this can fluctuate. Also, some groups find that multiple cores do appear in a proportion of showers containing 3×10^5 to several$\times10^6$ particles. The possible significance of this is referred to later. In the single-cored showers at least, a large part of the energy may be carried by one nuclear-interacting particle, found within < 1 m of the axis.

Further out, the muons arise from the decay of charged π-mesons, and the reason for their large spread is that the charged π-mesons of very high energy rarely decay, as the relativistic time-dilation increases their lifetime: instead they generate further π-mesons by colliding with air nuclei, until the particle energies are only a few GeV, when they decay; and at these energies the angles of emission are quite large.

6.2 The development of an air shower

Since the particle density ϱ has a simple distribution, it is possible from samples of the density to evaluate the total number N of electrons (positive plus negative) in a shower. It is clear that the size N of the shower at sea level produced by an incident primary particle of energy E_p will depend on the details of the nuclear cascade process which initiates the shower, if we wish to improve on Auger's rough estimate, and it is uncertain to the extent to which the rate of development of the shower in the upper atmosphere is not known. An approximate picture of the development of showers can be obtained, however, from the observations made in 1962–6 by an American–Bolivian–Japanese collaborative experiment at an altitude of 5200 m in the Andes near La Paz, just below Mt. Chacaltaya, just half-way down through the atmosphere (520 g cm^{-2}). Showers were recorded arriving at different angles ϑ to the zenith, and hence through different amounts of air (520 sec ϑ g cm^{-2}). If all showers of size N at depth t in the atmosphere follow the same law of development and thus have equal sizes N' at another depth t', the function $N'(t')$ can be mapped out by noting that if the rate of occurrence of showers of size $\geqslant N$ at t is R (e.g. 10^{-7} m^{-2} s^{-1} sr^{-1}), then this will also be the rate of showers $\geqslant N'$ at t'.

Figure 23 shows the number of electrons in showers observed under different amounts of air (a) for showers which arrive at a rate 10^{-7} m^{-2} s^{-1} sr^{-1}, and (b) for those arriving at a rate 10^{-10} m^{-2} s^{-1} sr^{-1}. The shading shows the possible extrapolations to smaller depths of air suggested by different cascade models. As the particles are relativistic and losing 2·2 MeV per g cm^{-2} (N being derived from scintillators on this basis), the area under the curves, as far as an atmospheric depth of 1030 g cm^{-2}, gives the energy loss by ionization in the normal atmosphere. Adding to this the energy still present in such showers at sea level, the total energy of showers of these particular sizes may be obtained, as shown in Table 5.1.

The authors of the work quoted (LaPointe *et al.*, 1968) using a detailed cascade model obtained energies 11 % higher and 1 % lower in the

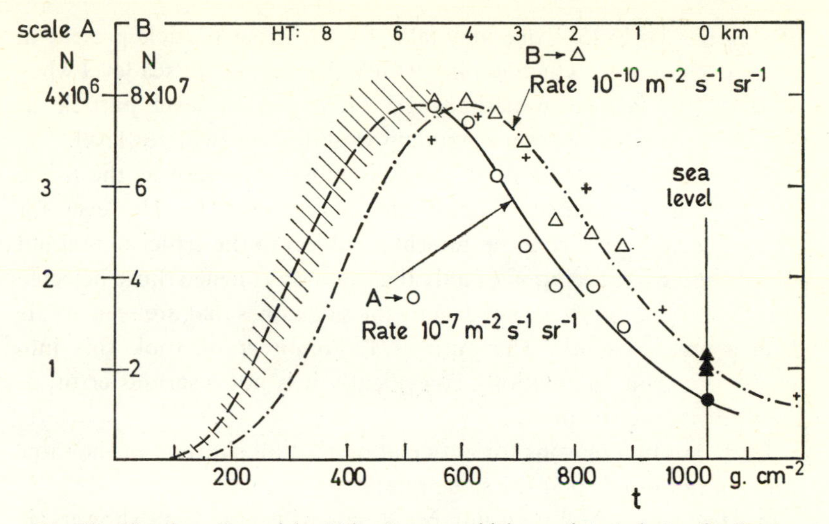

FIG. 23. Number of particles N in showers which are detected at a fixed rate (A or B) under different thicknesses t of air, from inclined showers at Chacaltaya, and from sea-level laboratories. $N(t)$ is believed to show approximately the growth and decay of showers of a particular energy. One relies on theory to continue the curve in the upper half of the atmosphere. (The crosses are obtained by scaling down data from a later Chacaltaya publication.)

TABLE 5.1.

ENERGY CONTENT OF SHOWERS OF TWO SELECTED SIZES

Size N at shower maximum	$3{\cdot}9 \times 10^{6}$	$7{\cdot}8 \times 10^{7}$
Size N at sea level	$6{\cdot}3 \times 10^{5}$	$2{\cdot}0 \times 10^{7}$
Ionization above sea level	$4{\cdot}4 \times 10^{15}$ eV	$8{\cdot}8 \times 10^{16}$ eV
Energy of soft component at sea level	$0{\cdot}14 \times 10^{15}$	$0{\cdot}4 \times 10^{16}$
Energy of nucleons and πs at sea level	$0{\cdot}08 \times 10^{15}$	0·2?
Energy of muons at sea level	$0{\cdot}64 \times 10^{15}$	$1{\cdot}3 \times 10^{16}$
Energy of neutrinos (estimated) at sea level	$0{\cdot}33 \times 10^{15}$	$0{\cdot}6 \times 10^{16}$
Total energy E_p	$5{\cdot}6 \times 10^{15}$ eV	$11{\cdot}3 \times 10^{16}$ eV

two cases. To $\sim 25\%$ one may take 1·4 GeV per particle present at shower maximum; which implies 9 GeV per particle at sea level when $N \sim 6 \times 10^5$, falling to perhaps 1·5 GeV per sea-level particle at $N \sim 10^{11}$ when the shower maximum would occur near sea level.

This conversion from N to energy is most important as the prime aim of the experiments is to detect high-energy particles. However, for the smaller shower size the argument is subject to the criticism that not all showers will develop in exactly the same way; hence those detected at rate R at one depth are not exactly the same ones that are seen at rate R at a greater depth. The authors LaPointe *et al.* took this into account in their calculations, so evidently it is not a serious error, as their results are so similar.

There are two reasons for anticipating that all showers of the same size will not develop at the same rate:

(a) Heavy nuclei, of mass number A, say, will cause some showers: it is to be expected that these will consist of A superimposed showers each of energy E/A, from the separate nucleons, and shower maximum will occur somewhat higher in the atmosphere than for a proton of energy E.

(b) The distance travelled by a proton between nuclear collisions will fluctuate about the mean of 80 g cm^{-2}; and if the whole shower development were to be displaced downwards by 100 g cm^{-2}, for example, the proton would require only 60% of the normal energy to produce a shower of given size at sea level, and the greater abundance of less energetic particles in the spectrum (by a factor $\sim 2·5$, say) would to some extent offset the improbability of the fluctuation. Fluctuations in the "age" of showers may thus be the hallmark of protons.

6.3 The energy spectrum above 10^{14} eV

The magnetic fields in interstellar space deflect cosmic-ray particles and obscure their sources, and apparently trap particles inside the Galaxy for a long period; hence the great interest in particles in the highest energy range, which may at some point cease to be trapped or greatly deflected. The maximum energy to which particles are accelerated is also relevant to their possible origin. It is therefore of interest to give a short account of the results obtained subsequent to Auger's work.

Only 1 particle above 10^{16} eV arrives per m^2 sr per year. The experiments mentioned earlier have detected showers of $> 10^7$ particles, but to go much further requires a collecting area of the order of square kilometres. Such systems have been set up, but the detectors have to be a long way apart, and can only sample the particle density at a few points in a shower, usually rather far from the axis. An early system was the ingenious array of Geiger counters of Cranshaw and Galbraith (1954, 1957), but this did not locate the axes of the smaller showers very accurately. The MIT group has been most extensively involved, first in experiments near sea level (Clark *et al.*, 1961), and later at higher altitudes. Very large plastic scintillation counters (up to 3·3 m^2) were deployed in concentric rings, which not only sampled the particle density at several points (from which the core position could be deduced, to about 10 m, and the size found), but the direction of arrival could be found from the delays in the times of impact on the different detectors. From Figure 22 it is seen that with such detector areas a signal is detectable up to hundreds of metres from the shower axis, depending on shower size.

As these experiments revealed no clear anisotropy in the radiation, and no sign of a falling-off in the shower size spectrum beyond 10^8 particles, larger arrays have been built—at Volcano Ranch (an MIT offshoot, at 1800 m altitude, in New Mexico) covering 8 km^2, and later at Haverah Park (260 m, in England) covering 12 km^2. One is faced here with the problem that in some showers there may be no measurement of particle density within 500 m of the axis, whereas half the shower particles are probably within 40 m of the axis, and various attempts to determine the total shower size or energy may differ. In presenting the collected results in Figure 24, consistent forms of the lateral distribution were used to perform this extrapolation and assign a shower size: this gives for the Volcano Ranch showers sizes nearly twice as large as those published (Linsley, 1963), so for the very largest showers the figures require confirmation. Preliminary Haverah Park

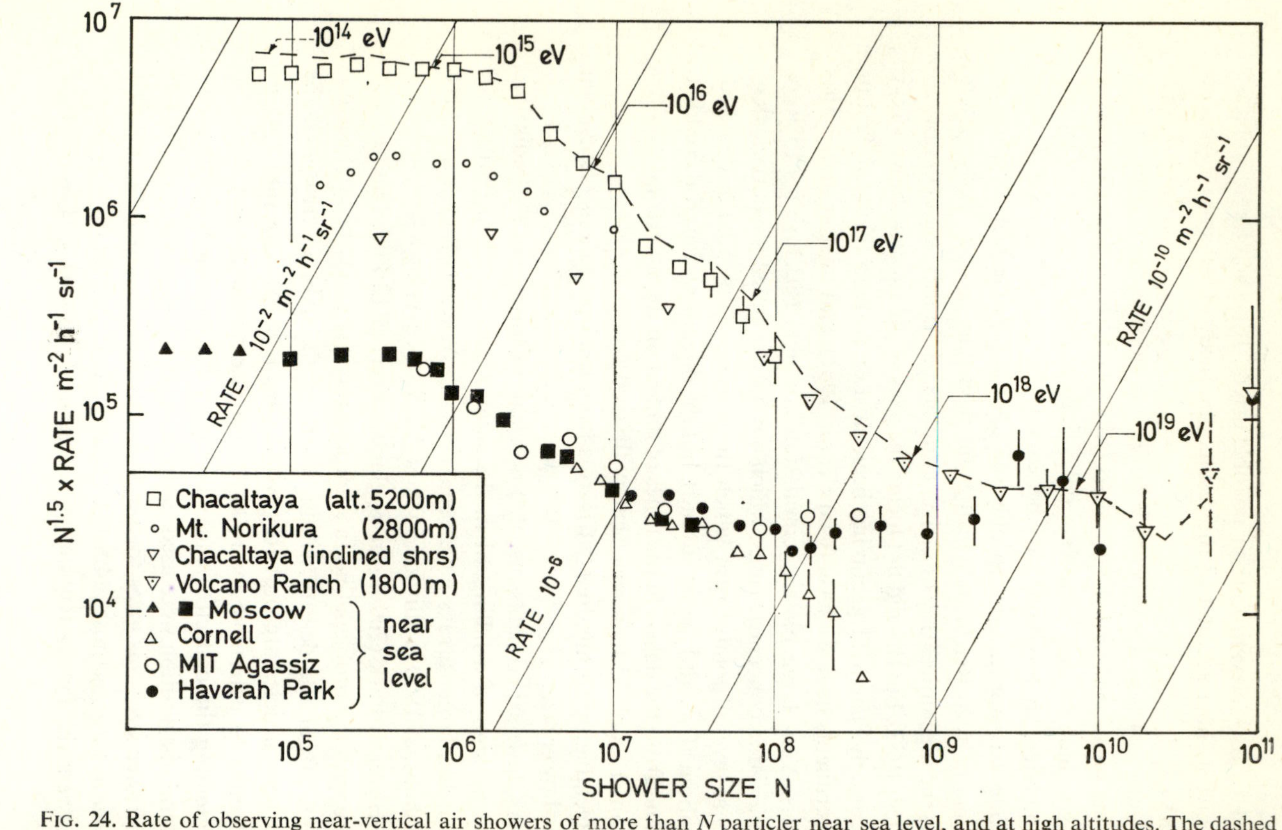

FIG. 24. Rate of observing near-vertical air showers of more than N particler near sea level, and at high altitudes. The dashed line represents the envelope of all such curves. The "inclined" Chacaltaya showess shown have passed through the same amount of air as the vertical showers at Volcano Ranch.

results in this range were plotted after deriving the shower size N from initial measurements which were not simply of the density of charged particles. (I am indebted to Dr. A. A. Watson for the basic data.) The results of the smaller array at Chacaltaya are also plotted: they are particularly valuable for $N \sim 10^{6 \cdot 5}$ to 10^8, as these showers are near their maximum development at that altitude (see Figure 23); thus the primary energy is more simply related to the shower size (almost proportional, and insensitive to fluctuations in depth).

Again, let R be the rate of arrival of showers of more than N particles at the observing level. It is convenient to present the integral shower-size spectrum $R(N)$ by plotting $N^{1 \cdot 5}R(N)$ in Figure 24. For small showers, $R \propto N^{-1 \cdot 5}$, so the graph is horizontal. To obtain the energy spectrum of the primary cosmic rays we proceed in two stages. First, the envelope of the curves $R(N)$ at the various altitudes gives very nearly (i.e. apart from the effects of extreme fluctuations) the rate of showers having more than N particles at shower maximum. Then, taking $E = kN_{max}$, where k is not far from 1·4 GeV (within 20%, say) over the region of interest, one merely has to relabel the abscissa in terms of energy, rather than N. Representing the integral energy spectrum as $F(E)$, the line on Figure 24 then gives $(E/k)^{1 \cdot 5}F(E)$.

A change in the slope of the spectrum is clearly shown, at a size 5×10^5 at sea level, and at $\sim 1 \cdot 8 \times 10^6$ at 520 g cm^{-2}, corresponding to a primary energy of about 3×10^{15} eV. The sharp break was first noticed by the Russian groups (Kulikov and Khristiansen, 1958), and the most plausible interpretation associates it with the onset of escape of particles from some local trapping region, probably the Galaxy. In the next few years one may expect further attempts to verify this. Peters (1960) pointed out that if it were indeed the curvature of trajectories in magnetic fields that determined the local retention of particles, the spectra might fall sharply beyond a certain magnetic rigidity; for heavy nuclei of charge z this would correspond to an energy z times higher than for protons; so just beyond the "knee" of the energy spectrum the primary radiation should be much richer in heavy nuclei, assuming the composition at the sources is not energy dependent. Several investigators have indeed claimed to see signs of this in an increased frequency, in showers of $N \sim 10^6$, of multiple cores, or multiple energetic particles, interpreted as remnants of the break-up of a large nucleus (particularly McCusker's group: Bray *et al.*, 1964; Dobrotin *et al.*, 1958, observed similar effects), though the details are in dispute at present, as different techniques give different results. Alternatively, the start of some energy loss process in the accelerating region or subsequently may perhaps be found to cause the change in slope.

Above 10^{18} eV the spectrum seems to regain a smaller slope. Noting that in a 5–10 μgauss uniform field a 10^{19} eV proton would have a 2–1 kpc radius of curvature, one can hardly escape the conclusion that these very energetic particles move freely through intergalactic space, a component of the total cosmic ray flux with a spectral index of $\gamma \sim 2{\cdot}6$ when we return to differential energy spectra, as is found at lower energies, but lower by a factor $\lesssim 100$ than the main flux (see Figure 25).

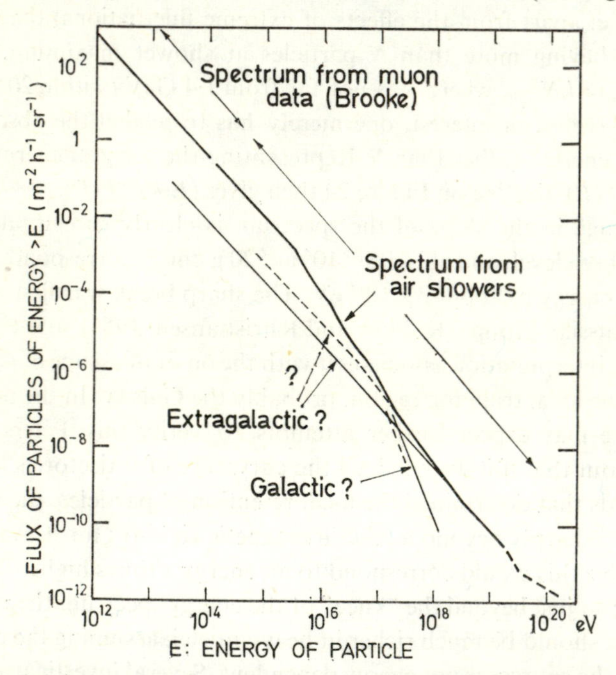

FIG. 25. One interpretation of the energy spectrum of primary particles as the sum of galactic and extragalactic components.

This background intensity could be even higher (e.g. dashed line in Figure 25), for the flux may be mainly intergalactic protons beyond 2×10^{17} eV according to Linsley and Scarsi (1962), and the sea-level N spectrum hints at an upturn about here, as though a more penetrating primary particle is appearing again. This, however, presents another puzzle. If the most energetic particles are protons, and have been produced much more than 10^8 years ago, the spectrum should start to drop appreciably by 10^{20} eV because of interactions with photons (section 9.1), and four showers with energy best estimated as between 1 and 2×10^{20} eV have been reported. Heavier nuclei should survive even less well. This is right at the limit of present observations, however, and the difficulties involved may be appreciated by looking at the range of shower rates written on the diagonal lines in Figure 24. A giant air shower array now coming into operation near Sydney and intended to cover 250 km² should be capable of advancing the known spectrum a further 0·5–0·8 decade.

The discussion of the energy spectrum has centred around the source and mode of transport of cosmic rays; more directly relevant perhaps are the direction of arrival of showers. From 1948 onwards very many investigators have sought anisotropy with the tantalizing result that small but apparently statistically significant effects have frequently been reported, only to disappear with further data. In many experiments the directions of arrival of individual showers were not measurable, but for most showers atmospheric absorption restricts the effective zone of sensitivity to a cone of about 0·7 steradian about the vertical, so that the Earth's rotation sweeps a band of the sky past the detector. The counts may then be Fourier analysed as a function of sidereal time; and if n showers are detected, a purely random distribution would by chance give a Fourier amplitude of $\sqrt{(2/n)}$ (r.m.s.): on this basis the amplitude of the first harmonic is less than 0·1 % at 3×10^{14} eV, $<0{\cdot}2\%$ at $1-2 \times 10^{15}$ eV, $< 1\frac{1}{2}\%$ at 10^{16} eV, $< 2\%$ at 10^{17} eV, and $13 \pm 6\%$ (hardly significant) at 2×10^{18} eV. On the previous interpretation of the energy spectrum, one might perhaps expect signs of escape from the Galaxy somewhere around 3×10^{16} eV, which is a region covered by only one experiment of high statistical weight—the first one by Cranshaw and Galbraith—in which apparently very significant 2%–5% amplitudes were found, although their later experiment at Culham showed no such effect at an only slightly higher energy. In view of the variations amongst the amplitudes quoted, the anisotropies which have appeared are not yet very convincing, but surprisingly there is a general tendency to agree on the time of maximum—around 19 to 20 h sidereal time (seen mostly from the northern hemisphere) for $< 4\times 10^{15}$ eV, and around 10 to 12 h for 4×10^{15} to 10^{17} eV, according to a summary by Sakakibara (1965). This dividing line in energy would seem significant, and the time of maximum for the lower energies would roughly correspond to particles diffusing along the (outward) direction of a galactic spiral arm.

VII

The Origin of Cosmic Rays

THE central problem of cosmic ray physics is, of course, how and where the particles acquired their immense energies. The high total energy density must derive from a major energy source, and the large energies of individual particles require explanation.

Millikan's idea of direct production in nuclear reactions fails on the latter count, apart from other reasons. Lemaître made an intriguing suggestion that cosmic rays were the debris of the original disintegration of the "primeval atom", a super-nucleus from which the expansion of the universe was supposed to have begun, 10^{10} years ago. Heavy nuclei present a problem here, since at a very early stage when the universe was small and dense they would have suffered collisions and disintegration. It now seems that the radiation from this primeval system would have been in the form of γ-rays, which have suffered an enormous red-shift during the expansion and are now detectable as far infrared photons.

The estimates of the age of cosmic rays which have already been quoted suggest that we look for a generation mechanism still at work.

7.1 The possibility of electromagnetic acceleration

When the latitude effect had indicated that cosmic rays were charged particles with typical energies $\sim 10^{10}$ eV, Swann (1933) made the first plausible suggestion of how cosmic-ray energies might be attained—reprinted as Paper 11 in Part 2—showing that electrons might be accelerated to such energies by the e.m.f. induced by changing magnetic fields near the surface of the Sun and stars; for it had been known since 1908 that magnetic fields of up to several kilogauss are associated with sunspots, which may appear and disperse over a period of days or weeks on the Sun's surface.

The situation which Swann discusses is a simplification of the field of a sunspot: the magnetic field lines emerge normally to the surface over a large area, and there is an axis of symmetry, where the field is strongest. The whole field increases gradually in strength. In a numerical example, an average field of 4000 gauss inside a circle of radius 3×10^5 km is supposed to build up in 10^6 sec: this rate of change of flux induces an e.m.f. of $1{\cdot}1\times10^{11}$ volts around the circle. This radius is nearly half the radius of the Sun, and is much too large for a sunspot, but Swann considered that such spots might occur on other stars. It seems from this that suitably large energies might be imparted to electrons or ions, but the general motion of a charged particle in such a changing electric and magnetic field is complex, depending on the way in which $\boldsymbol{B}$ varies with radius, r, so Swann uses a Hamiltonian formalism for the problem, as it yields certain simplifying constants of the motion in a very symmetrical situation.

For readers unfamiliar with this formalism, a very brief summary will be given.

If the only field sources are currents, the scalar potential may be taken as zero, and the field represented by the vector potential $\boldsymbol{A}$ (Swann uses $\boldsymbol{U}$):

In MKS units, $\boldsymbol{E} = -\partial \boldsymbol{A}/\partial t$, $\boldsymbol{B} = \text{curl } \boldsymbol{A}$, $\boldsymbol{A}$ being directed in circles in the plane normal to $\boldsymbol{B}$, for the form of field assumed. Given the field, the motion of the particle in r, ϑ, z (cylindrical polar coordinates) may be deduced from a Lagrangian function L of r, ϑ, z, and $\dot{r}$, $\dot{\vartheta}$, $\dot{z}$, the appropriate Lagrangian being given by Swann. The classical momenta P_r, P_ϑ, P_z, conjugate to the coordinates in the equations of motion are derived by partial differentiation of $L(r, \vartheta, z, \dot{r}, \dot{\vartheta}, \dot{z})$.

Thus $$P_r = \partial L(r, \vartheta, z, \dot{r}, \dot{\vartheta}, \dot{z})/\partial \dot{r}, \ldots, \text{etc.}$$

The Hamiltonian function H is given from L and the momenta by Swann's equation (4). Its magnitude is found to be equal to the particle's energy (kinetic+rest energy) in this formulation (to be distinguished from an alternative covariant Hamiltonian often used in similar problems), and when H is expressed as a function of $r, \vartheta, z, P_r, P_\vartheta, P_z$ (not $\dot{r}, \dot{\vartheta}$, etc.), the equations of motion are

$$P_r = -\partial H/\partial r, \; r = \partial H/\partial P_r \quad \text{(and similarly for } \vartheta, z\text{)}.$$

Note (Swann's equations 1 to 3) that these generalized momenta do not simply represent the kinetic momentum $\boldsymbol{p} = m\boldsymbol{v}$ of the particle. Another property of this function $H(r, \ldots, P_r, \ldots)$ is that the total rate of change of its value as the particle moves, dH/dt, is equal to $\partial H/\partial t$.

The special example described by Swann had the interesting property of permitting continued cyclic acceleration; an electron starts from rest by being accelerated by $\boldsymbol{E}$ along a circumferential direction, but is then deflected by $\boldsymbol{B}$, and with B assumed $\propto r^{-1}$, the radius of the motion remains constant. This special situation, in which the average $\dot{B}$ inside the orbit is twice the value of $\dot{B}$ at the orbit, is the condition for the betatron accelerator, which was later successfully constructed (1941) to accelerate electrons, but seems artificial for stellar regions. Swann did not insist on cyclic acceleration, but with the more realistic assumption of acceleration along only part of an orbit, giving 2×10^{10} eV in the example, it was not proved that the electron retains its energy on escaping.

Another difficulty is that such a field system is a form of transformer, requiring primary currents flowing in the same region; so why are only a few of the charges present accelerated to relativistic energies? Though later work by Riddiford and Butler indicated that even without the special form of $\boldsymbol{B}$ sunspot fields might accelerate particles very effectively, Parker (1958a) showed that in these sunspot models (including Swann's) the behaviour of the growing field in the first few milliseconds is vital to the setting up of the orbit; in a conducting gas the fields could not grow from zero in the ideal manner assumed, and acceleration is negligible.

Provided that the fractional change in B is not large in the period of the Larmor orbit, and B does not vary too suddenly in space, $p_\perp^2/B$ will remain constant, where $p_\perp$ is the component of the (kinetic) momentum of the particle normal to $\boldsymbol{B}$, whilst the parallel component $p_{||}$ is not altered. For a single cycle of growth of field, this gives much less acceleration.

The importance of Swann's suggestion of betatron acceleration was in showing that changing cosmic magnetic fields might accelerate particles to sufficient energies, rather than in the detailed arrangement discussed.

There are, however, stars which have large-scale magnetic fields in the kilogauss range, rather than gauss, and the possibility that these vary in such a way as to make large-scale acceleration possible cannot be ruled out. For instance Alfvén has considered the possibility that two magnetic stars rotate around each other, as binary stars are common, to produce an electric field along the axis of rotation, but the energy of revolution must be rapidly damped if such systems are truly effective generators. Babcock found that some magnetic stars can even reverse their polarity in a matter of days, and proposed that the associated electric fields may be important particle accelerators. However, amongst magnetic stars, the most promising candidates may turn out to be the recently discovered pulsars; for according to the most plausible current explanation these are rapidly rotating magnetic neutron stars (Gold, 1968), formed by the collapse to tens of km radius of highly evolved stars. If the magnetic flux is conserved during the collapse, surface field strengths could be 10^{10}–10^{12} gauss. (Both rotational and magnetic energy are increased by shrinkage.) If the magnetic and rotational axes differed, very powerful electromagnetic fields would be available, but it is too early to assess the various models proposed for generation of the pulsed radio emission.

7.2 Fermi's acceleration mechanism

The debate on the means by which cosmic rays acquire their energy turned in a new direction after a paper by Fermi—"On the origin of the cosmic radiation"—published in 1949, and reprinted as Paper 12 in Part 2. With the possible exception of the poorly understood magnetic variable stars, the likely cosmic ray output from stellar fields had seemed low, whereas Fermi's proposed mechanism—the collision of fast-moving charged particles with moving magnetic fields—made available the kinetic energy of large-scale turbulent motion in the interstellar gas as an energy source. Randomly moving clouds of gas were supposed to carry with them relatively intense rigidly attached magnetic fields (as described in section 5.3) from which charged particles would bounce elastically, if they could not penetrate. In an attractively simple analysis of the process, Fermi treats this as a simple collision problem: in the frame of reference of the cloud the particle has no change of energy, but transforming to the galactic coordinates the particle gains energy in a head-on collision or loses energy in an overtaking collision, the latter happening slightly less often. The result is on average a fractional increase α per collision in the particle's total energy (including rest energy). For clouds moving with r.m.s. velocity Bc, $\alpha \sim B^2$ (perhaps up to $3B^2$ in Fermi's equation 14). With the additional assumption that the particle may at any time be lost from the system (nuclear collision being considered) Fermi shows that the particles will build up an energy spectrum

$$J(E)\,dE \propto E^{-\gamma}\,dE, \quad \text{where} \quad \gamma = l+\frac{T_G}{T_L},$$

T_L being the loss lifetime and T_G^{-1} being the average fractional energy gain per unit time: $T_G^{-1} = \alpha/\tau$, if τ is the mean time between collisions with the clouds. This spectrum simply reflects the chance of experiencing a particular number of accelerating collisions before the particle is lost.

The attractions of Fermi's account are that continued acceleration, though slow, does not depend on idealized forms of fields, that a power-law spectrum appears a natural result, provided T_L and τ are constants, the upper limit of energy acquired is limited only by the requirement that the strength and extent of the moving field suffice to turn back the particle and by the time available, and that the mechanism will not squander its energy on all ions in space, but will add more energy to those few which have already become energetic. A major difficulty, which rendered the model incomplete, was that below a certain threshold energy the acceleration would not balance energy losses by ionization: another source must supply protons with at least 200 MeV; and more for heavy nuclei.

Two fatal objections were soon raised to the model in its original form, however. In the following year observations on primary nuclei began to show that cosmic rays did not remain in circulation until they suffered nuclear collision (see the paper by Bradt and Peters), but had lived in the galactic disc for only 2 or 3 million years; so T_L was much shorter than Fermi assumed. Moreover the spectra of different nuclei were similar, unlike the expectation if T_L depended on nuclear cross-sections. And Unsöld (1951) estimated that the velocities of moving clouds which might be encountered very frequently were nearer 3 km/s than the 30 km/s which Fermi took: thus T_G would be 100 times greater and the spectrum 100 times steeper. Taken together, these points indicated that a much more rapid acceleration mechanism was required.

In 1954 Fermi suggested a somewhat different approach, when these problems were apparent, noting that observations of a polarization imposed on starlight had recently been interpreted as showing that the magnetic field in galactic spiral arms was more regular than previously assumed, pointing along the arms (and aligning paramagnetic dust particles in space). Cosmic rays, constrained to circle around guiding centres moving along the field lines, might then bounce back and forth between approaching regions of higher magnetic intensity, giving a series of head-on reflections with acceleration $\alpha \sim 2B$ rather than B^2 per collision, until the particle escaped from the trap. (For example, we remember that $p_\perp^2 \propto B$, so $\sin^2 \vartheta \propto B$, and the particle is reflected if it moves into a field such that $\sin \vartheta$ reaches 1, and if B is 50% above normal, 58% of an originally isotropic distribution of particles will be reflected, but will recoil with increased $p_{||}$, so ϑ will be less at the next encounter with the stronger field, and the trap will eventually be penetrated after an energy gain of say 10%. Sharp bends in field lines might scatter the particles to regain isotropy of ϑ, so further acceleration could occur; though one must admit appreciable deceleration between receding fields.)

In the spirit of Fermi's simple approach, it is helpful to picture the cosmic-ray particles as atoms of a gas, which gain energy by collisions with a hotter gas of much heavier molecules—the gas clouds (Fermi's analogy)—or, in the modified scheme, by compression between approaching (leaky) pistons. Even leaky pistons will do work on the gas as there is always a higher gas pressure ahead than behind: a sequence of irreversible expansions and compressions heats the gas (Parker, 1958a). For one-dimensional compression of relativistic particles, the energy per particle is readily shown to be inversely proportional to the distance between the pistons (i.e. to the volume, V), energy density varying as V^{-n} where $n = 2$; though $n = \frac{4}{3}$ if frequent scattering maintains isotropy. One sees that compressions of $\geqslant$ 10-fold are required to accelerate through each decade of energy, neglecting the deceleration (or expansion) phase, so that if acceleration of this type is important one must envisage many successive phases of compression: for the pistons one would need criss-crossing trains of magneto-hydrodynamic waves. The time constant T_G for this acceleration, before correction for expansion phases, is of the order of the time taken for waves to meet. (To give $T_G \sim 5\times10^6$ y, for $\gamma = 1{\cdot}6$, might require regions where reflecting waves $\sim$40 pc apart were moving at 10 km s^{-1}.)

Fermi also perceived that the cosmic radiation must affect the dynamics of the interstellar gas, and that some kind of dynamic equilibrium between cosmic rays, magnetic energy, and turbulent kinetic energy (respective energy densities W_{cr}, W_m, W_t) may exist—an idea which has persisted. Extending ideas of Pikel'ner and Shklovskii, Parker (1958a) proposed a balancing process in which magnetic waves in the moving gas compressed the cosmic-ray "gas", which reacted back by modifying the velocity of sound, $\sim\sqrt{(\text{pressure}/\varrho)}$, if its own pressure, $P_{cr} \doteqdot \frac{1}{3} W_{cr}$, were added to the other pressures, and he pointed to several instances of disturbances moving faster than the velocity of sound in interstellar gas (1·6 km s^{-1} at 100°K). But more recent calculations on wave propagation in the interstellar medium indicate that the magnetic waves which are required should be rapidly damped. Without very small-scale field irregularities to scatter the particles, the cosmic-ray gas will prove too "slippery" to compress, and the short waves required are likely to die out rapidly away from active stellar sources.

The close similarity of W_{cr} ($\sim$1 eV cm^{-3}), W_m (1 eV cm^{-3} for 6·3 μgauss, near what is required to fit radio emission), and turbulent energy, also suggested that the gas motions were imparting their energy to cosmic rays (or the reverse!), but in view of the difficulties in maintaining the special forms of waves needed for acceleration, a different basis of equilibrium may be sought. In his more recent work, for instance, Parker proposes that the magnetic field acts as a safety valve on the pressure P_{cr}. As the magnetic field in the disc has probably a rather regular pattern, an origin in galactic rotation rather than entirely in turbulent motions is suggested, and as in this model the field lines run through the gas parallel to the disc plane, the cosmic rays cannot escape along the field, so their pressure tends to inflate the disc, leading to $P_{cr} \simeq W_m$ (Parker, 1969): at the disc boundary the lower field cannot withstand P_{cr}, and is blown out, giving us a model of the halo as made up of bubbles of ejected field, gas and particles. It is seen that the cosmic ray gas is being given a prominent role in galactic dynamics—it may even influence condensation of gas clouds—but this is not bringing us much closer to the acceleration of cosmic rays.

Ginzburg (1953) pointed out that conditions would be much more favourable for Fermi acceleration in regions where velocities of moving plasmas were higher and collision intervals shorter, as in disturbed regions of the solar atmosphere (accelerating possibly to 10^9 eV) and in supernova remnants, such as the Crab nebula. Small highly-disturbed regions might thus accelerate particles. If Fermi's process operates right from the start, it will give more energy to nuclei than to electrons, in proportion to their rest energies.

Where there are moving magnetic fields there may also be fields varying cyclically in strength, where betatron acceleration could also take place, for $p_{\perp}^2$ would increase with B, and if the radiation were then made isotropic by scattering from magnetic irregularities, one-third of the increase in $p_{\perp}^2$ would be given to $p_{||}^2$, which is unaltered by induction, and thus not all of the energy gain would be lost when B fell to its original level. Such gains could be of comparable importance to the others discussed.

However, one unsatisfying feature of Fermi's theory of the energy spectrum is that it makes γ depend sensitively on the rates of acceleration and loss of the particles, which must vary in different systems, whereas different galaxies show evidence of very similar spectra. One has to conclude that the equilibrium between cosmic rays and plasma maintains $T_G = 1{\cdot}6T_L$. Syrovatskii (1961) suggested that this arose from an equilibrium in which $W_{cr} = W_m = W_t$ all the time in the source region, whilst the number of cosmic-ray particles was dwindling by leakage, there being no other energy loss, so that the individual particle energies rose, whence a simple calculation of the number leaking after the mean energy is E shows that $\gamma = 2{\cdot}5$ for those emitted. But the weakest fields then have to contain the most energetic particles.

Although the number of different mechanisms for acceleration in conducting media is limited, it is not yet exhausted. Another important one could be the rapid dissipation of magnetic energy by reconnection of magnetic lines of force near a neutral line or surface, if this occurs in a region of very low density. As usual, it is hard to know whether the required field configurations actually arise.

We have spent rather a long time on this inconclusive discussion, but the interactions of plasmas and relativistic particles are of great importance; and the complexities hidden in the laws of electromagnetism are surprising.

7.3 Supernovae as sources

We now turn from the hypothetical means by which energy might be imparted to particles to the identification of effective sources. Even before the radio evidence for concentrations of relativistic electrons in the Crab nebula, it had been noted that supernova explosions might provide sufficient energy to maintain the cosmic radiation if much of the output went into relativistic particles (ter Haar, 1950). In section 3 of his paper reprinted (no. 15) in Part 2, Ginzburg shows that the interpretation of the radio emission gives a firm basis for this assumption, an estimate of the total output of cosmic-ray energy by all supernovae matching within an order of magnitude or two the rate at which the Galaxy loses cosmic rays.

The nature of supernovae is still not well understood. The term nova means a "new" star, and is applied to the case where what appears at first sight to be a new star suddenly shines in the sky. In reality an old faint star has suddenly flared up to become immensely brighter. Supernovae form a distinct group of much more intense stellar outbursts, in which for several days a star becomes intrinsically $\sim 10^9$ to 10^{10} times as bright as a star like our own sun, and then fades away over a period of months. Such outbursts are rare, and during the era of the large astronomical telescopes none has been seen in our galaxy, but only in very distant galaxies. Many of those observed, including examples in our own galaxy observed visually in 1054, 1572 and 1604, have a characteristic variation of brightness with time, finally decaying with a 55-day "half-life": these are classified as "type I", and are apparently explosions from old and not very massive stars ($\sim$solar mass), which eject a shell of material (mass very much less than a solar mass) with velocities $> 10^3$ km s^{-1}, releasing up to 10^{50} ergs. There is at least one other type, however (type II), which though possibly less bright optically releases perhaps 10^{52} ergs in ejecting a much greater mass of material with velocity ~ 7000 km s^{-1}. These appear to have been very massive young stars (found only in spiral arms of galaxies): some evidence suggests that the star may be largely disrupted. Examples of remnants of type II are the radio source Cassiopeia A (not seen visually, though estimated to have exploded in 1700) and the Cygnus loop nebula.

According to theories of stellar evolution the explosion occurs in certain stars when the nuclear fuel in the core has been converted to iron, the most tightly bound nucleus, after which no further release of nuclear energy can support the weight of the star, and collapse occurs in minutes or seconds as endothermic reactions ensue. Various mechanisms for causing a "bounce" or ejection of an outer shell may exist. Gravitation is the source of the energy released. We now see expanding shells of turbulent gas at the sites of such outbursts.

Ginzburg's first estimate of output was necessarily based on rough estimates, applying the useful rule attributed to him, that in astrophysics, $10 \doteqdot 1$. One may check how well his conclusion stands up to later figures. Concerning the rate at which supernovae occur, Shklovskii (1960), making use of old Chinese records, found good evidence for 6 supernovae in our galaxy in 20 centuries, and as these apparently fell in a 30° sector of the galaxy and in an annulus of 5 kpc, the overall rate might have been 1 per 15 years—or say 40 years to a factor 2·5, as some distances may have been underestimated (cf. Ginzburg's 300 y, an estimate then current). If the present particle content of supernova remnants ($\sim 2\times 10^{49}$ erg from section 5.5) is what is eventually emptied into space, uniform production over the whole galactic disc would yield $10^{49\cdot 3\pm 0\cdot 3}$ erg/$10^{9\cdot 1\pm 0\cdot 4}$ s = $10^{40\cdot 2\pm 0\cdot 5}$ erg s^{-1}. From section 4.4 we saw that the cosmic rays apparently spent 3×10^{6} y in the gaseous galactic disc, or half this if the general density were 2 atoms cm^{-3} (though it could be less if the fragmentation occurred largely at the source), and spreading the local cosmic-ray density (0·8–1 eV cm^{-3}) over a gas volume $\pi\times(15\ \text{kpc})^2\times 0\cdot 3$ kpc (neglecting the central bulge, as the other figures refer to disc conditions) we require an energy input of $10^{-11\cdot 84\pm 0\cdot 05}$ erg $cm^{-3}\times 10^{66\cdot 8\pm 0\cdot 3}$ $cm^3/10^{13\cdot 8\pm 0\cdot 3}$ s = $10^{41\cdot 2\pm 0\cdot 5}$ erg s^{-1}. Thus, if the observations are rightly interpreted, supernovae can quite possibly supply the energy (at the worst, a few per cent of it). (Ginzburg, at least, is one of those people who know when to apply his rule.)

Unfortunately there is still no direct evidence of the flux of nuclei in supernovae; only the electrons radiate, and all the previous figures assume 100 times as much energy in nuclei as in electrons. γ-rays characteristic of meson production may some day be detected; but the only evidence at present is that the chemical composition of the primary radiation shows the source to be unusually rich in iron. (The alternative explanation that the primary acceleration process picks out heavier nuclei is unlikely, as (a) the emphasis seems to fall particularly in the iron region, and does not increase beyond iron, and (b) particles are occasionally accelerated in solar flares, and show a normal composition.) This would fit in well with debris from a star which has completed its nuclear evolution. The confirmation of trans-uranic elements would point strongly to such a source.

Regarding particle acceleration, Ginzburg suggests that energy is transferred from the turbulent elements of gas in the nebula by Fermi acceleration of the random type, referring to his earlier (1953) paper. The really uncertain factor is the scale of these elements: their effect must overcome the opposing effect of expansion. Repeated cycles of induction might also help here. But 10^{12} eV electrons pose a problem. The maximum proton energy contained might be $\sim 10^{16}$ eV, for in a 10^{-3} gauss field the diameter of its trajectory would be $\sim 1/100$ that of the Crab nebula.

Two other sites for the acceleration have been considered.

Colgate and his co-workers (e.g. Colgate and White, 1963) have examined a mathematical model of a type II supernova explosion, in which a very strong shock wave travels out after the explosion, driven initially by neutrinos, and blows off the star's outer layers, reaching relativistic velocity in the less dense regions. It is supposed that radiation pressure separates electrons and ions, giving a travelling electric wave which can accelerate the ions in the outer layer directly to give the correct cosmic-ray energy spectrum up to at least 10^{16} eV; though trans-uranic nuclei are not expected. Several steps still remain to be justified. Experimental evidence of the envisaged millisecond shock-wave would be interesting.

Oort and Walraven (1956) believed that remarkable moving ripples which recur every few months near the centre of the Crab nebula (Figure 21, p. 78) were the site of acceleration in magnetic shock waves: about 10^{38} erg s^{-1} may be involved. Their further suggestion that the star from which they recede is still active has now been vindicated most strikingly: after a pulsar with the remarkably short pulsation period of 33 ms was found in the direction of the Crab nebula, Cocke, Disney and Taylor (1969) showed that the central region was emitting light pulses of the same repetition frequency, the main ones lasting only 1·4 ms, and within weeks Miller and Wampler (1969) produced pictures taken through a rotating shutter synchronized to this period using the Lick 120 in. telescope, and showed that the star in question was responsible (see Figure 21), emitting most of its light in short flashes.

Shklovskii (1966) had estimated that 10^{38} erg s^{-1} is still being put into electrons in the Crab nebula, as those above 10^{12} eV are very short-lived. Now the pulsar is found to be slowing down, probably losing rotational energy by electromagnetic braking at a rate quoted as 10^{38} erg s^{-1}. It remains to be seen whether this is a coincidence.

After this note of optimism, we may remind ourselves that the Crab appears not to be such an efficient proton accelerator.

7.4 Other sources of cosmic rays

(a) Ginzburg suggests that ordinary novae may manufacture cosmic rays: as they each release $\sim 10^{-4}$ of the energy of a supernova, but occur at $\sim 100\,y^{-1}$, their total energy output may be a half that of supernovae, so perhaps their cosmic-ray output is in similar proportion.

(b) Acceleration may take place at the galactic centre, where the radio emission is stronger (though perhaps this only means the magnetic field is higher). In the strong extra-galactic radio sources, there is general evidence of violent emission of material from the nuclear region, as though explosive activity in galactic nuclei is not uncommon, and many normal galaxies show radio emission mainly from a central region. Lacking quantitative data on this aspect of our galaxy at present, we cannot assess this possibly very important source in the same way as supernovae.

(c) Burbidge and Hoyle (1964) have urged that cosmic rays may be largely of extra-galactic origin, arguments by Burbidge and counter-arguments by Ginzburg and Syrovatskii having continued at the successive international conferences. Of course, if cosmic rays leak out of our galaxy at 10^{41} erg s^{-1}, they presumably do likewise from most others, and with a density of average galaxies $\sim 10^{-75}$ cm^{-3} (Allen, 1963) leaking steadily for 10^{10} y, one only has from this source (allowing for red-shifts) an energy density $\sim 1{\cdot}5\times10^{-5}$ of that seen near the Earth. The strong radio-galaxies which may spill particles at 10^{3} times the rate of normal ones (if the relativistic particles escape in $\sim 10^{6}$ y) have a space density $\sim 2\times10^{-79}$ cm^{-3}, so may only double the rate. It is believed that weak magnetic fields exist even outside galaxies, and particles would hence follow irregular paths, wandering only a few tens of Mpc even at 10^{18} eV according to Ginzburg and

Syrovatskii; so one may consider the supposed "local supergalaxy" to be the limit of the region of interest, where the density of average galaxies may be at least 10 times higher. One may then guess an extragalactic flux $\sim 3\times10^{-4}$ of the observed general flux.

The air shower spectrum above 10^{18} eV may be taken to indicate an extragalactic flux $\geqslant \frac{1}{100}$ of the galactic flux, in which case a factor $\geqslant 30$ is to be accounted for. Radio source counts have indicated much greater activity in the past than at present, corresponding to a very high output when galaxies were young (e.g. Longair, 1966), and hence such a factor might be retrieved, if we take it that these extra-galactic rays are very old. But then interactions with the supposed far infrared (3°K) photons in intergalactic space should have reduced the flux appreciably above 10^{18} eV and very greatly above 5×10^{19} eV. So the position is not yet clear.

(d) One gains the impression from Figure 15 (p. 61) that there is a separate component of cosmic rays below 30 MeV, with a very steep spectrum. Since at 2–3 MeV the range of a proton is $< 0{\cdot}01$ g cm^{-2}, their source is probably fairly local, perhaps some other kind of star. Or, the solar system may simply admit slow particles more readily.

7.5 Motion and storage of cosmic rays

A by-product of Fermi's discussion of random collisions with gas clouds, which remained even after the efficacy of the acceleration process came into doubt, is the idea of a random walk of cosmic-ray particles in the Galaxy as the means of producing a storage time long compared with the time for a direct escape path: nearly random changes of direction are assumed at intervals of tens of parsecs. Seen on a large scale, the motion is like diffusion. Ginzburg and Syrovatskii have developed much further the model which Ginzburg originally put forward. In their model, the particles diffuse through the near-spherical halo, with a scattering mean free path $\lambda \sim 10$ pc, giving a diffusion coefficient $D = \frac{1}{3}\lambda v \sim 10^{29}$ cm^2 s^{-1} (v, the velocity of the guiding centre, being $\leqslant 1{\cdot}5\times10^{10}$ cm s^{-1}). Their sources are concentrated near the centre of the system, and the particles have a typical storage life $\sim 3\times10^8$ y, mostly spent in the halo, with some chance diffusion through the disc. (The same assumed diffusion coefficient would give an escape time of only $\sim x^2/2D \sim 10^5$ y if particles moving out of the disc, of thickness x, were lost freely.) The assumption of free circulation of particles from the halo also appeared necessary when 10^{18} eV particles were first detected, as the fields in the galactic disc clearly could not retain them (the radius of curvature for a proton of 10^{18} eV moving normal to a field of b microgauss being $1{\cdot}0/b$ kpc), but this difficulty eased with the discovery that the spectrum dropped at $\sim 3\times10^{15}$ eV.

At the end of his paper Ginzburg underlined the importance of measuring the electron component of cosmic rays; and now this has been done we can see how the new data narrow down the possible models, raising some difficulties for the halo model in fact. From equation (1) of Alfvén and Herlofson's paper (p. 254), it follows that electrons lose energy at a rate $\propto E^2$ by synchrotron radiation, and that in a field whose component normal to the motion is $b_\perp$ μgauss, there will be no particles left with energy above $E_c = 8400\ \text{GeV}/(b_\perp^2 T)$ after T million years. If the electrons were injected continuously for T million years, with a spectrum $E^{-\gamma}\,dE$, the spectrum would appear steepened to $E^{-\gamma-1}\,dE$ for $E > E_c$. If the electrons spent most of their time in the halo, $E_c \sim 1{\cdot}7$ GeV (assuming $T \sim 300$, $b_\perp \sim 4$: really $b_\perp \sim 2\frac{1}{2}$, but another energy loss process due to scattering by 3°K photons may be represented by adding $(2{\cdot}8)^2$ to $b_\perp^2$). If the disc is the containment region ($T \lesssim 2$, $b_\perp \sim 6$), $E_c \gtrsim 117$ GeV. The spectrum shown in Figure 17, on p. 68, after allowing for solar modulation at energies 0·5–3 GeV, has $\gamma \doteqdot 2{\cdot}5$ up to ~ 100 GeV at least, pointing to the latter situation (Anand, Daniel and Stephens, 1968): the alternative interpretation that $E_c \sim 0{\cdot}5$ GeV requires $T \sim 1000$, and also gives $\gamma+1 \doteqdot 2{\cdot}5$, so that the initial electron spectrum is very different from the nuclear spectra, despite their striking actual similarity. (Consideration of energy losses similarly rules out altogether the possibility that the electron component is of extra-galactic origin.) Clearly there must be particles in the halo, as they escape from the disc; the question is whether they return significantly to the disc. In either disc or halo model, the particles reaching us would make most of their nuclear collisions in the disc (the halo has only $\sim 10^{-2}$ atoms cm^{-3}), so the amount of fragmentation presumably gives us the time spent in the disc, as we assumed when estimating the power requirements.

Finally, the diffusion of the cosmic rays is clearly related to their near isotropy. If the flux observed in directions ϑ from some axis is $\propto (1+A\cos\vartheta)$, it appears that $A \lesssim 0{\cdot}1\%$. If the local particles are isotropic within some system, but this system is moving with a drift velocity v, then transformation of the particle fluxes and energies into our coordinate frame gives $A = (\gamma+2)v/c$ if the energies are relativistic. Hence $v \lesssim 70$ km s^{-1}: the particles are clearly carried round by the galactic rotation. If they have diffused a distance r from a point source in time t, then $v \sim r/2t$, and a source at the galactic centre would indeed require $t > 70$ million years to make A so small, as in the halo storage model. But in a more disc-like model, a magnetic field predominantly running along the disc would guide the diffusion almost entirely along this direction, and if the effective sources are in the disc around us the radial flow may be much altered, particularly if the magnetically guided diffusion paths spill a part of the radiation out of the disc at many places (as in Parker, 1969). Theoretical work now indicates that magnetic waves excited in the local plasma by cosmic rays may conspire to destroy an isotropic flow velocity which much exceeds the velocity of Alfvén waves, and so the source directions may thus be heavily disguised.

It is thus possible to devise incomplete models for cosmic-ray production, which require more astronomical data to progress further. There is still more to be learned about the interaction of fast charged particles and plasmas.

VIII

Local Peculiarities

8.1 Solar modulation: the sunspot cycle

The discovery that the intensity of cosmic rays below 10 GeV varies considerably during the cycle of sunspot activity has been recounted in section 4.2. Although the abundance of radioactive by-products (e.g. C^{14}, Cl^{36}, Be^{10}) shows that the flux of cosmic rays has not changed much (certainly within a factor 2) on a time scale of 10^4 or 10^6 years, there are many short-term variations, whose causes must be sought locally, within the solar system. Until these are better understood we cannot know what is the general galactic flux of cosmic rays; but this is only one reason for studying them: another is their promise of illuminating many aspects of the interaction between fast particles and plasmas which may be relevant in a wider context. However, only in the last few years has a unifying picture of interplanetary space appeared, to provide a framework for discussion of these very complex effects. A short introduction to this picture will be given below, as a background to current research using space probes.

From the earliest days of cosmic ray investigations, time variations have claimed a great deal of attention. Observed variations became smaller as instrumental effects were reduced, and effects of atmospheric origin were recognized and corrected for. Precision ionization counters have been used, but at ground level they have to be shielded from variable local radiations; and large-area counters in coincidence can monitor the muon rates. The neutron monitor is now the standard rate monitoring device, introduced around 1949 by Simpson in Chicago and independently by Adams and Braddick in Manchester: a world-wide network had been set up by the International Geophysical Year in 1957. This instrument uses BF_3 counters to detect neutrons evaporated from nuclear interactions of neutrons and protons (and to a lesser extent pions and muons) in a large mass of lead. The long-term stability of the monitor can be checked and it has the advantage that the atmospheric conditions only affect the nucleonic particles through the barometric pressure, the rate decreasing by a factor e for an increase of 140 g cm^{-2} in the overlying air. The muon flux, by contrast, varies also with the temperature as the height of the 100 millibar layer largely determines the chance that a muon will have decayed. From latitude surveys, it is found that 50% of the counts in a sea-level neutron monitor come from primary particles with rigidities below 19 GV, 20% from rigidities below 7 GV, 5% below 4·7 GV and $\frac{1}{2}$% below 2 GV, for a primary spectrum obtaining at sunspot minimum (Quenby, 1967). Muon monitors respond to higher energies.

The main variations recognized by neutron monitors near sea level are:

(a) The decrease of up to 20% in rate, roughly as the monthly number of sunspots increases, following an 11-year cycle. Figure 26 shows the monthly counts of the Leeds neutron monitor since 1954, corrected for atmospheric effects, alongside sunspot counts. (The sensitivity of neutron monitors may have changed by up to 2% due to changes made during this period.)

(b) Forbush decreases—relatively sudden decreases in intensity (a few per cent up to 20%) occurring on a world-wide scale, followed by a slower recovery usually lasting very many days. The main part of the fall may sometimes take up to two days, but often occurs in 2 to 5 hr or even less. It probably always occurs within a short time of a "geomagnetic storm sudden commencement" when the intensity of the Earth's magnetic field shows a sudden small rise, presaging a highly disturbed period when the field strength falls.

(c) 27-day recurrences of variations—If one day shows an abnormal level of the intensity, a similar displacement may on average be found

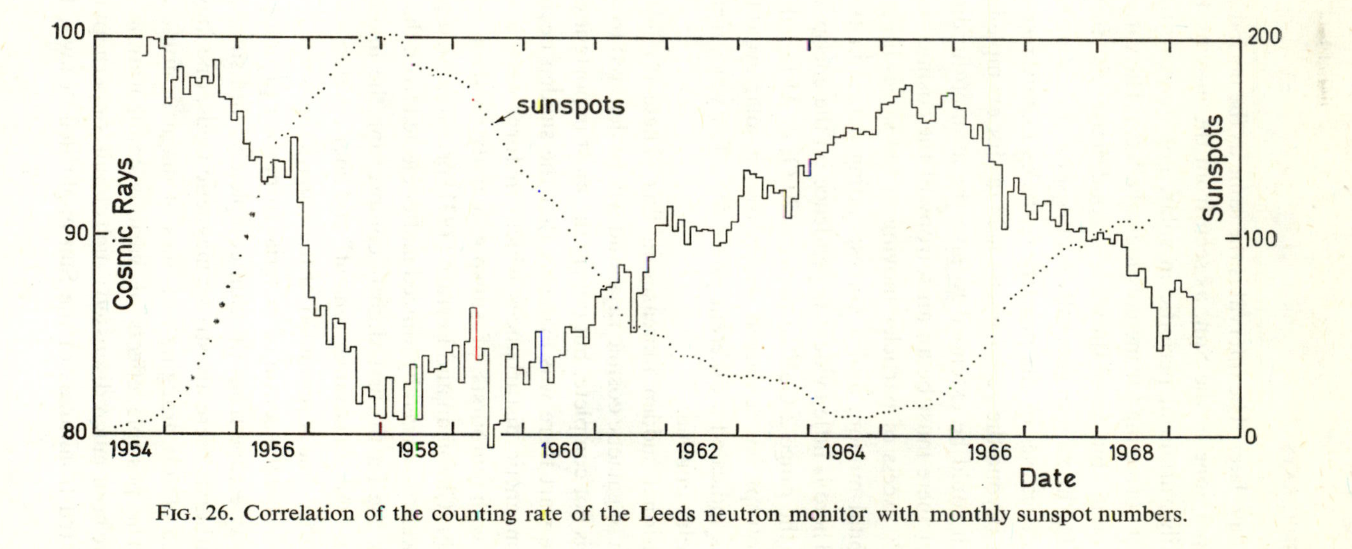

FIG. 26. Correlation of the counting rate of the Leeds neutron monitor with monthly sunspot numbers.

to recur 27 days later. The interval is the same as the period of rotation of the sunspot zone of the Sun, as seen from the moving Earth.

(d) A daily variation, periodic in solar, not sidereal, time, which remains after correcting for meteorological effects. The variation is not quite sinusoidal, but also shows a second harmonic (semi-diurnal variation). The maximum generally occurs soon after noon, local time; but if the trajectories of the primary particles mainly responsible are traced back beyond the geomagnetic field (as for example in Figure 7, p. 22, which could be extended to give the asymptotic directions) it appears that there must be an anisotropy of the cosmic-ray flux with a 0·4 – 0·3% excess of particles moving in the same direction as the Earth's orbital motion, and a corresponding deficit in the opposite direction. There is little evident dependence on the energy of the particles, over the range 2 GeV to perhaps 70 GeV. Although the phase of the anisotropy has remained steady over a solar cycle from 1957, earlier work indicated a different phase for a few years near the 1933 and 1954 solar minima.

(e) Occasional sudden increases in flux. These do not represent modulation of galactic cosmic rays, and will be deferred to section 8.2.

This list is not complete, but covers the most important effects.

From the start there was evidence that the sudden decreases were related to magnetic disturbances which, in turn, were known to be correlated with outbursts of sunspot activity, with a propagation delay $\sim$ 30 hr. Chapman and Ferraro (1931) proposed that geomagnetic storms were produced by ionized matter ejected from the Sun which later struck the Earth, first slightly compressing the magnetosphere, producing a "sudden commencement" and later enveloping it. Following this further, one is reminded of the remarkable solar photographs which show violent ejection of material from solar prominences, some of which may escape from the Sun completely; and since at the sites of these outbursts there are strong magnetic fields, one may expect to have plasma clouds dragging out lines of magnetic force away from the Sun. The possible effects of various configurations of ejected plasma have been much discussed: such clouds, or perhaps continuous beams emitted from areas of the Sun might deflect away low energy cosmic-ray particles as required for a Forbush decrease, and a long-lived beam might engulf the Earth every 27 days, as the Sun rotates.

However, a very large-scale effect is required to explain the 11-year variation. Much of the recent work in this field derives from the idea of the solar wind, developed by Parker (1958b, c, 1963), and first sketched in the final paper reprinted in Part 2, entitled "Cosmic ray modulation by solar wind", in which Parker suggests that a gaseous wind constantly blowing out from the Sun, but more strongly and effectively at times of high solar activity, can push back low energy cosmic-ray particles entering the solar system from outside.

He refers to two earlier sources for the basic idea of the solar wind —to Biermann who had, a few years earlier, postulated such an outflow to account for the acceleration of comets' tails—and to his own earlier paper (Parker, 1958b) setting out a simple dynamical basis for the phenomenon by showing that the observed temperature of the solar corona, about $1{\cdot}5 \times 10^{6}$°K, was too high to permit a static gravitational equilibrium: instead, the gas pressure, maintained by the source of heating, must accelerate the outer layers to a remarkable extent, reaching supersonic velocities. This basis for the model, from Parker's earlier paper, will be outlined below, before looking at some of its other consequences.

We suppose that there is spherical symmetry, and at a radial distance r, there are n hydrogen ions and n electrons per cm^3 making up a gas flowing out with velocity V, the temperature being assumed constant at T, $\sim 1{\cdot}5 \times 10^{6}$°K. The equation of continuity of flux is

$$nVr^2 = n_0V_0r_0^2, \tag{8.1}$$

where the suffix refers to some arbitrary reference layer. The gas pressure $2nkT$ has a gradient which accelerates the gas, so

$$mnV\frac{dV}{dr} = -\frac{d}{dr}(2nkT) - \frac{GMmn}{r},$$

where M, m, refer to the masses of the Sun and a proton. d/dt has been replaced by Vd/dr for conditions of steady outflow,

i.e.
$$\frac{d}{dr}\left(\frac{1}{2}mV^2 + 2kT\ln n - \frac{GMm}{r}\right) = 0. \tag{8.2}$$

Writing $\frac{1}{2}mV^2 = \varepsilon$, $GMm/r = \gamma$, $kT = \Theta$, using (8.1) to substitute for $\ln n$, and noting that $r/r_0 = \gamma_0/\gamma$, we can write the equation connecting the three forms of particle energy as

$$(\varepsilon-\varepsilon_0)-\Theta \ln\left(\frac{\varepsilon}{\varepsilon_0}\right) = 4\Theta \ln\left(\frac{\gamma}{\gamma_0}\right)+(\gamma-\gamma_0). \qquad (8.3)$$

This passes through a critical point at a distance r_c where each side of equation (8.3) reaches a minimum. Here, it is easily found that $\varepsilon = \Theta = \gamma/4$; i.e. the energy of bulk motion equals the thermal energy kT. We take this as our reference distance, r_0, so that the equation simplifies to

$$\frac{\varepsilon}{\Theta}-\ln\left(\frac{\varepsilon}{\Theta}\right) = 4\ln\left(\frac{4\Theta}{\gamma}\right)-3+\frac{\gamma}{\Theta}. \qquad (8.4)$$

At radial distances much less than r_0, the bulk motion is very small; beyond r_0 it has been accelerated to supersonic velocity. This can only happen if r_0 is greater than the Sun's radius, 7×10^5 km (i.e. near the solar surface the gravitational energy $\gamma > 4kT$), otherwise ions would have to be injected with supersonic velocities to maintain a steady flow.

If $T \sim 1{\cdot}5\times10^6$°K, the thermal velocity $\sqrt{(2kT/m)}$ is 158 km s^{-1}, and the escape velocity $\sqrt{(2\gamma/m)}$ at 7×10^5 km is 620 km s^{-1}. Hence if the corona maintains its temperature to a distance of $\sim$0·1 A.U.† ($1{\cdot}5\times10^7$ km), one finds from (8.4) that $\varepsilon = 4{\cdot}56\Theta$, corresponding to an outward velocity there of 340 km s^{-1}, which is highly supersonic. If the further expansion becomes adiabatic, the eventual velocity reached (which turns out to be roughly that found in isothermal conditions at $2\frac{1}{2}$ times this transition distance) is 500 km s^{-1}. Near the solar surface, however, V is only 0·5 km s^{-1}, easily supplied by the surface gas. (Strictly, though, one should not take the base of the corona right at the solar surface.) Whilst Parker's first isothermal model is only a rough approximation, in reality some mechanism does heat the corona to maintain a nearly constant temperature well beyond the critical dis-

† The astronomical unit of distance (A.U.) is the semi-major axis of the Earth's orbit—$1{\cdot}5\times10^8$ km.

tance (3×10^6 km at this temperature); and in 1962 the Mariner 2 space probe confirmed this model by detecting an outflowing stream of protons with velocity $\sim$500 km s^{-1} (Neugebauer and Snyder, 1962), though further frequent monitoring shows that the velocity is not steady: the "normal" velocity near the Earth's orbit may be around 400 km s^{-1}, but temporary increases often occur, up to double this value. A typical density is $\lesssim$10 ions cm^{-3}. When increased solar activity raises the coronal temperature, an increase of wind velocity and density is expected.

As the conductivity of the ionized gas will be very high, the magnetic field relaxation time will be much longer than a solar cycle, so the outflowing gas will drag magnetic lines of force with it (cf. section 5.3); and thus in Parker's expanding corona one expects an Archimedean spiral form of magnetic field (Figure 27) as a particular line of force

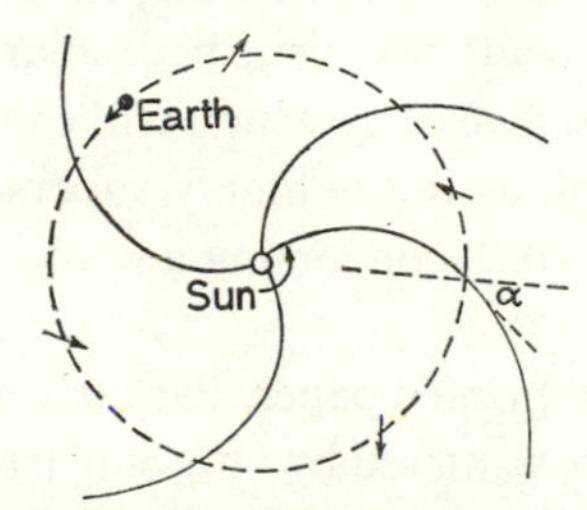

Fig. 27. Spiral form of interplanetary field lines in the ecliptic plane according to Parker. There are found to be sectors in which the field direction is alternately inward and outward, as shown.

threads through the portions of gas successively emitted from the rotating Sun, rather like the pattern of water from a rotating garden sprinkler. (The energy density of the field is too small to affect the mass motion of the gas and prevent this.) At a radial distance r the line of force lags roughly by an angle $\varphi = \omega r/V$ behind its azimuthal position at the Sun (if one approximates the solar wind velocity as a constant), where ω is the Sun's angular velocity ($2{\cdot}9\times10^{-6}$ s^{-1}); and the lines make an angle $\alpha = \tan^{-1}\varphi$ with the radius vector, or 45° near the Earth's orbit if $V = 430$ km s^{-1}. The magnitude of the field on this basis, expressed by its radial and tangential components, should be:

$$B_r = B_s\left(\frac{r_s}{r}\right)^2, \quad B_\varphi = B_s\left(\frac{\omega r_s}{V}\right)\left(\frac{r_s}{r}\right)\sin\vartheta,$$

where the polar angle ϑ is 90° in the plane of the Earth's orbit (the ecliptic), and r_s is the solar radius, B_s is the field strength there, at the point where the line leaves the Sun. Taking the observed fields in the Sun's photosphere to be typically 1 gauss, 30 μG is the predicted strength near the Earth. From 1963, the IMP (interplanetary monitoring platform) satellites with highly elongated orbits have confirmed both the direction and magnitude quoted above for the average local interplanetary magnetic field, though with appreciable sudden disturbances at times. According to the observations, the sense of the field reverses at intervals around the Sun, with positive and negative sectors as indicated in Figure 27. It was not unexpected that sharp irregularities are found in the field; for disturbed solar regions may project plasma with an increased velocity which will overtake the normal field and form a shock front, as in the highly supersonic flow information cannot be passed back to the oncoming gas when the wind encounters any obstacle.

Coming now to the reprinted paper, Parker's suggestion was that the cosmic-ray particles are scattered by outgoing irregularities in the field; they have to diffuse in through an outflowing medium to reach the Earth. The natural supposition that there are more irregularities, and also that V is greater, when the Sun is more disturbed leads to the correlation of cosmic-ray intensity with the sunspot cycle; and the change in shape of spectrum which he presents looks very reasonable.

In this early paper, Parker did not make use of the spiral form of the field lines, but further development has shown that there should be additional observable effects of the rotation, as will be shown.

The high conductivity of the plasma implies that $\boldsymbol{E}$ must be very small in a co-moving frame of reference, so that in a stationary frame $\boldsymbol{E} = \boldsymbol{V}\times\boldsymbol{B} = VB\sin\alpha$ normal to the ecliptic. This field will cause a drift, with velocity $-\boldsymbol{E}\times\boldsymbol{B}/B^2$ to be superimposed on the normal motion of the guiding centre of a cosmic-ray proton's helical motion along a magnetic field line. This drift is in the ecliptic plane, perpendicular to the field lines, having a radially outward component $V \sin^2 \alpha$, and a tangential component $V \sin \alpha \cos \alpha$. The analysis is complicated by the gradient in B which causes another drift normal to the ecliptic, and which in a steady and regular field configuration would set up a normal density gradient whose pressure would exactly offset the force and the drift due to the electric field. In the real field it seems that this gradient is much reduced, partly by diffusion in a more disorderly field shell further out, and partly by the frequent reversals of the field direction. However, the components of the cosmic ray electric drift velocity imposed by the wind are partly offset, becoming $fV \sin^2 \alpha$ and $fV \sin \alpha \cos \alpha$, where $f \sim \frac{1}{2}$. Neglecting particle production and absorption by the Sun itself, one expects no net radial flux, so it must be supposed that a radial density gradient of particles causes an inward diffusion, as considered by Parker, with a diffusion coefficient D along the field lines, and much smaller diffusion across the field. Denoting the cosmic-ray particle density by n, the condition for the radial component of convection to balance that of diffusion is

$$nfV \sin^2 \alpha = D \frac{dn}{dr} \cos^2 \alpha. \tag{8.5}$$

But if convection and diffusion are in balance radially, their tangential components add, giving a total drift velocity w parallel to the Earth's orbital motion: $w = fV \tan \alpha = f\omega r$—i.e. a fraction f of what would correspond to a rigid rotation of the "cosmic-ray gas" with the Sun. If $f \sim 0{\cdot}6$, this is just what is required to explain the main component of the diurnal variation.

Returning to the density of cosmic-ray particles, the reduction below the density far away from the Sun is, by integrating (8.5),

$$n = n_\infty \exp\left(-\int_r^\infty \frac{fV}{D} \tan^2 \alpha \, dr\right).$$

Parker's original expression is similar to this of course, though α did not appear, and his result may be written more simply than in his paper as the following function of rigidity R:

$$n = n_\infty \exp(-\eta/\beta R), \tag{8.6}$$

where βc is the particle velocity, and η is independent of the particle's rigidity. This form has also been proposed by Dorman (1963) to fit observations and the Parker model. The result depends on the details of the scattering process—how it depends on particle energy. With Parker's assumptions about the scattering path length, the diffusion coefficient $D \propto \beta R$; and recent space measurements of the Fourier spectrum of field irregularities have justified a dependence of this form (Gloeckler and Jokipii, 1966), although for R somewhat below 1 GV this relation breaks down. Only partial quantitative success has been achieved to date in describing the modulated spectra, which is not surprising in view of the possible complexities of the interplanetary field near the limit of the solar wind—a limit which may lie at 20 A.U. or perhaps much further—and hence the absolute value of the reduction in galactic flux (linked to the value of η) is not known; but it is clear that the wind still blows, and presumably lowers the flux, even at solar minimum. Various estimates suggest that $0{\cdot}6 \text{ GV} \leqslant \eta \leqslant 1{\cdot}0 \text{ GV}$, when modulation is at a minimum, and equation (8.5) was used, with these limits, to estimate the unmodulated fluxes in Figures 15 and 17, on pages 61 and 68.

An essential feature of the whole picture is a radial gradient in the density of cosmic rays in the solar system. This has been detected by the Mariner 2 Venus probe and the Mariner 4 flight to Mars, the latter finding that η in expression (8.5) decreased by 0·2 GV as r increased from 1 to 1·4 A.U. (for particles with $R > 0{\cdot}5$ GV, for which 8.5 appears valid).

Finally, a Forbush decrease must correspond to some form of disjointed field sweeping out from the Sun, and sweeping away cosmic rays; but its detailed morphology is not yet clear.

It is prudent to recall that this interpretation of cosmic-ray modulation is based on extrapolation of interplanetary conditions close to the Earth's orbit: what happens away from the ecliptic, and at greater distances may be very important. Also, particles probably lose energy in the moving fields; and this may account for $\sim \frac{1}{3}$ of the modulation.

8.2 Solar flares

Less than once a year the Sun generates, for a few minutes, relativistic particles which reach the surface of the Earth. The first instances recognized were those reported by Lange and Forbush (1946) as sudden increases in the level of ionization. Since 1949, when neutron monitors have been in use, the increases have been more prominent and have been found to follow, usually within a few minutes, the observation of great solar eruptions known as class 3^+ solar flares. Flares are observed as patches of bright emission in H_α light, lasting a few minutes, and occurring above the solar surface near sunspot regions. A few of the most intense have radiated white light, and given rise to radiofrequency outbursts, probably synchrotron radiation.

The cosmic ray effect is variable in the extreme: most solar flares generate no particles which reach ground level detectors; but on the other hand one of the earlier flares caused a monitor to be switched off in the belief that an electrical fault had developed. The largest sea-level flux increase so far observed is shown in Figure 28, the peak nucleon intensity at Leeds probably reaching 45 times normal after the great flare of Feb. 23, 1956. The flare was seen in Tokyo and Kodaikanal to reach its maximum brightness at 0342 h UT, probably 10 min after its start. Favourably placed stations recorded the first particles arriving at 0344, with a maximum intensity reached 4 min later. About this time the increase started at many other stations, reaching maximum about half an hour after the flare maximum.

If the particles all came from one direction—e.g. directly from the Sun—they would be directed magnetically to certain "impact zones" on the Earth. However, after the peak the arrival directions appear to be isotropic, and the two stations shown come into line (though Chicago has a slightly lower threshold rigidity), so there must be an appreciable amount of scattering, for motion along the diverging lines of the interplanetary field would tend to reduce the pitch angles of the helical trajectories. The variation with time of the detected flux has been examined with the hope of learning how the particles propagate, assuming they are generated in one short burst. Uniform isotropic

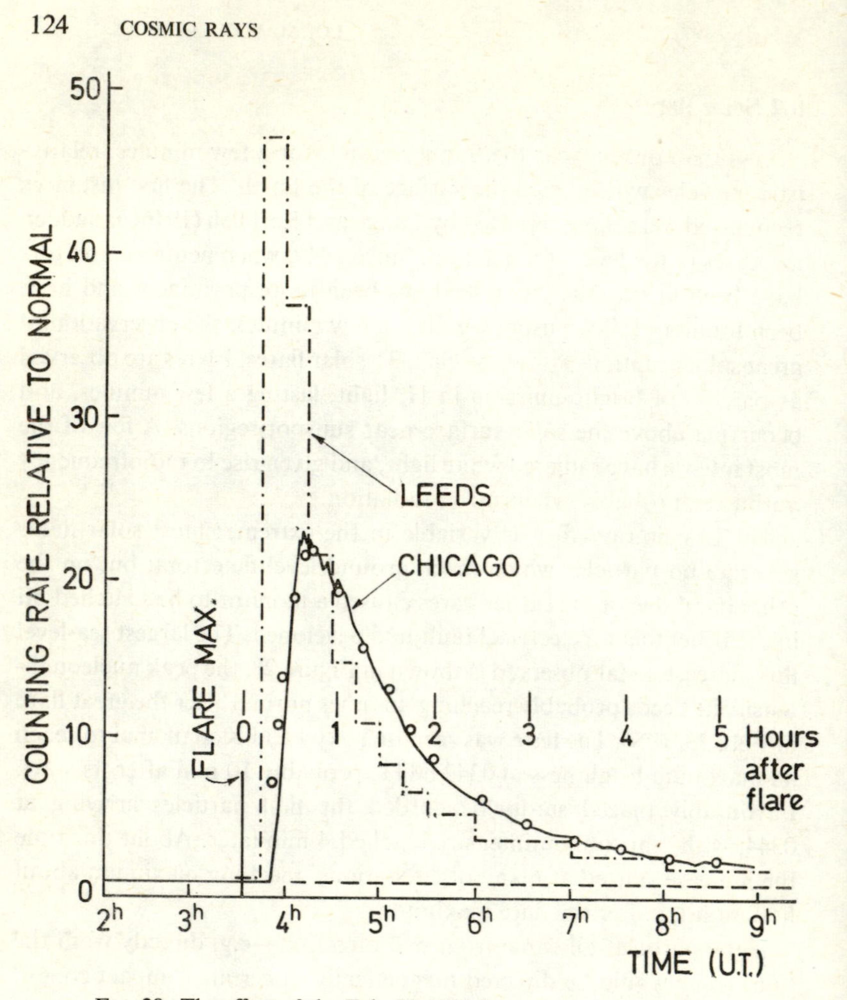

FIG. 28. The effect of the Feb. 23 1956 flare on neutron monitors.

diffusion would yield a particle density at a distance r, at the Earth, $\propto t^{-\frac{3}{2}} \exp(-kr^2/t)$, but $t^{-\frac{5}{2}} \exp(-f(r)/t)$ seems to fit better (as shown for the Chicago data in Figure 28 by the circles), as though the scale length for diffusion normal to the interplanetary field lines is very small but increases $\propto r$ (hence $\dot{\propto} \sqrt{t}$).

From the effect of later flares on monitors in different locations, McCracken (1962) was able to show that the first particles came from a direction ~ 50° west of the Sun, and combining this with the observation that cosmic ray increases are more frequently detected from flares on the Sun's western hemisphere, he derived strong arguments for the spiral form of the interplanetary field to guide particles along such a path, before detailed spacecraft observations were available. However, in some cases the normal field may have been disturbed by a previous outburst. Although the particles are probably largely confined to helices within expanding magnetic tubes, those of ~ 1 GeV are evidently rapidly scattered in pitch angle, so McCracken had to suppose there were many field irregularities of a size comparable to their gyromagnetic radius, which would have this effect. This model is consistent with that used to explain the 11-year variation.

From observations with balloons released shortly after flares began it is clear that the energy spectrum of the flare particles is much steeper than for galactic cosmic rays: differential rigidity spectra R^{-4} to R^{-7} have been derived; in other cases $\exp(-R/R_0)$, with $R_0 \sim 0{\cdot}04$ to 0·4 GV has given a better description. Hence sea-level detectors record a very small part of the true flux and indeed fail to see most flare events. (The limit of detection is around 1·1 GV.)

Since the International Geophysical Year, protons down to 30 MeV have been monitored by the increase in ionospheric opacity they cause at higher latitudes ("polar cap absorption"), revealing many more low-energy flare events; and in the last few years continuous monitoring of 3–500 MeV protons from spacecraft has completely overturned the idea that acceleration of protons in solar flares is rare: only the attainment of energies of several GeV is rare. These most sensitive detectors have not yet covered a period of maximum solar activity, but it seems that the integrated flux of solar protons above 20 MeV over a solar cycle must be an order of magnitude greater than the flux of 20 MeV galactic protons which reach the vicinity of the Earth, though at 1 GeV the solar flux is over two orders of magnitude below the galactic flux. The propagation of these low-energy protons is not yet understood, as sharp spatial variations occur, which have suggested a more intricate magnetic field pattern—a "spaghetti model", in which adjacent magnetic lines rapidly diffuse apart, and interweave.

Ginzburg (1953) and Parker (1957) have argued that the Fermi mechanism could be responsible for the acceleration of particles in flares, and the energy spectra are compatible with this: magnetic energy does in any case seem the likely source of the total energy release, and fields of several hundred gauss may be available. Other acceleration processes involving the destruction of magnetic flux have been considered, but observations differ as to whether a flare results in a change of field. Our ignorance of the conditions in the flare still leaves the method of acceleration open to speculation, and the determination of the mechanism would be of great interest in the wider context of the origin of cosmic rays. It is interesting in this connection to note that emulsion observations on heavy nuclei present in the solar radiation show that the acceleration process is quite unselective with regard to nuclear charge: the relative abundances from helium upwards at a given rigidity are very similar to the abundances believed to hold for the material of the solar photosphere.

8.3 The Earth's magnetosphere

Recent work has shown that the simple picture of the motion of charged particles in the geomagnetic field given in Chapter II is incomplete: there are other local cosmic ray effects associated with the magnetosphere which deserve a mention.

(i) The Earth's dipole field has a sharp boundary where it gives way to the interplanetary field. Pressure from the plasma of the solar wind deforms it so that some lines of force are blown out as a long tail behind the Earth, and on the sunward side the regular geomagnetic field ends sharply at about 10 Earth radii, at the magnetopause: beyond this lies a disturbed region where the solar plasma flows round the magnetosphere from which it is largely excluded; and near 14 Earth radii lies a magnetohydrodynamic collisionless shock wave marking the boundary of the steady solar wind. Figure 29 shows the field configuration constructed from numerous Russian and American satellite observations.

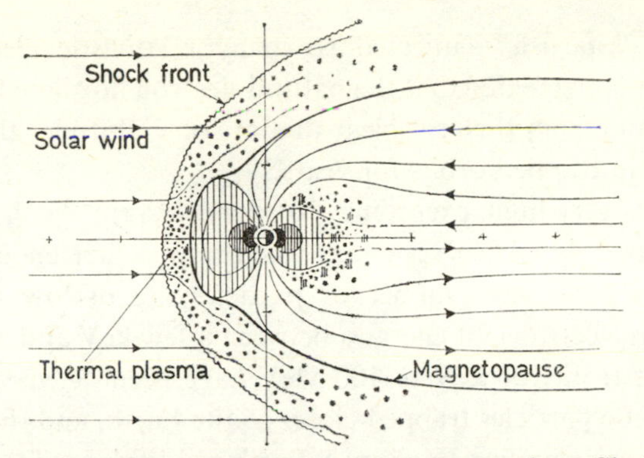

FIG. 29. Comet-like deformation of the outer magnetosphere. Shading marks areas of trapped radiation.

At high magnetic latitudes ($\gtrsim 60°$) the cut-off rigidity varies from day to night, e.g. from $\sim$ 160 MV to $\sim$ 15 MV at Churchill.

(ii) The presence of local and temporary islands of electrons of tens of keV—well downstream in the tail, near the magnetosphere boundaries, and precipitating in the auroral zones—shows that processes of particle acceleration take place in the magnetosphere, particularly in geomagnetically disturbed conditions. Theoretical work suggests also that there may be an electric field capable of accelerating electrons and protons by many tens of keV along the magnetic neutral plane which is found in the equatorial region of the tail. Hence one may have here another laboratory for studying particle acceleration by plasmas.

(iii) Belts of trapped radiation in the magnetosphere fill orbits which Størmer's theory (Chapter II) shows to be inaccessible to particles approaching from outside, and from which particles conversely cannot escape—for instance, from the inner allowed zone in Figure 4c on p. 18. The characteristic orbit is one in which a particle spirals around a line of the dipole field, with the pitch angle of the helix increasing as it approaches the Earth, $\sin^2 \vartheta/B$ remaining constant until when $\vartheta = \pi/2$ the particle turns back and spirals to another mirror point in the opposite hemisphere. A drift around the Earth is superimposed on this to-and-fro motion. In Figure 7 (p. 22), the particle of lowest rigidity is in a nearly-trapped orbit. In 1962, a high-altitude hydrogen bomb explosion—"Starfish"—injected so many relativistic electrons into the lower trapping region that the natural electron flux here has become obscured; and from this it is clear that at some altitudes the particles can remain in trapped orbits for years.

The flux is very high, exceeding 10^8 particles $cm^{-2}\,s^{-1}$, and completely stopped Van Allen's counters when it was first encountered in 1958; though the most numerous particles are of low penetrating power. Both electrons, of energies between a few keV and a few MeV, and protons from 100 keV to 100 MeV have been detected, with the more energetic particles trapped closer to the Earth, and the less energetic electrons extending to about 8 Earth radii. Most of the energy is in the low-energy protons.

Some of the inner-belt particles are injected by the decay of neutrons ejected from cosmic-ray interactions in the atmosphere, but the variability of the outer regions, and to a lesser extent the inner zone, requires a more rapid source of replenishment, and it appears that thermal protons from the solar plasma must be caught up in the varying outer fields and somehow gradually driven further in, increasing their energy as they pass into the stronger fields.

IX

Other Cosmic Radiations

HAVING seen in previous chapters the value of correlating various types of radiation, we end with a brief mention of some new possibilities which are opening up.

After the Millikan era, γ-rays were not believed to play a significant part in the primary radiation. Now, strenuous attempts are being made to detect them, and to relate to the charged cosmic rays other photons not coming from obvious stellar sources.

9.1 The universal 3°K photons

Dicke and his co-workers (1965) set out to detect traces of a black-body radiation which should still uniformly fill space if the universe had expanded from a very hot condensed state. Looking into the extreme distance ($\sim 10^{10}$ light years) in any direction, and hence into the distant past, the hot primeval material should be seen, but the large red-shift should move the radiation to the far infrared and microwave region, where it should still appear as a black-body radiation spectrum, cooled to a few °K. Penzias and Wilson (1965) discovered the radiation before the Princeton group, and many observations since have shown that the general radio background spectrum above a few hundred MHz, instead of falling as $\nu^{-0.7}$, suddenly rises as ν^2, and is consistent with a black-body radiation at 3°K, though confirmation of the spectrum above its expected peak frequency is still lacking. Even if the generally accepted cosmological explanation is incorrect, the remarkable isotropy indicates that it must extend far into extragalactic space, since space is very transparent to it. The energy density is $4{\cdot}8\times10^{-3}\,T^4$ eV cm^{-3} = 0·39 eV cm^{-3} if $T = 3°$.

In intergalactic space, high-energy cosmic rays are more liable to interact with this medium of photons than with matter, and so it may serve to make visible the presence of the rays, just as magnetic fields did within the Galaxy. In the rest frame of a cosmic-ray particle of mass M and energy $E = \gamma Mc^2$, such a photon of energy ε (mean: $2{\cdot}7kT = 7\times10^{-4}$ eV) would appear to have an energy $\bar{\varepsilon} = \gamma\varepsilon(1-\cos\vartheta)$, if $\gamma \gg 1$, where ϑ is the angle between the trajectories of particle and photon. Thus for protons above 10^{20} eV photoproduction of π-mesons can occur (Greisen, 1966), giving an energy-loss lifetime of $\sim 10^8$ y for the protons. (Above 2×10^{18} eV electron–pair production would occur sufficiently to give an energy-loss lifetime $< 10^{10}$ y; and these effects would have been more serious when the universe was smaller.) If such protons do come from extragalactic sources one thus expects a sharp drop in the energy spectrum above $\sim 5\times10^{19}$ eV.

The effect on electrons has already been mentioned (section 7.5), the rate of energy loss due to scattering on such photons being equivalent to that by synchrotron radiation in a magnetic field whose energy density is half that of the black-body radiation (i.e., $B_\perp = 2{\cdot}8$ μgauss).

9.2 Cosmic X-rays and γ-rays

Since 1962 rocket and balloon experiments have shown that there are localized strong sources of X-rays in the Galaxy, some possibly being nova remnants, and also a diffuse X-ray background, the latter having a flux $\sim 150E^{-2\cdot3}$ photons m^{-2} s^{-1} sr^{-1} MeV^{-1}, where E is in MeV, between 35 keV and 1 MeV (and $\sim 1350E^{-1\cdot68}$ in the same units for $0{\cdot}002 < E < 0{\cdot}035$ MeV); and there is a very strong flux below 1 keV. Much of this has not yet been interpreted.

Hoyle (1965) suggested that these diffuse X-rays were originally 3°K photons, which had been scattered by the relativistic cosmic-ray electrons; for the photon, which seems to the electron to have an energy $\bar{\varepsilon}$, may be scattered out at an angle $\bar{\varphi}$ in the electron's rest system (measured from the electron's direction of motion) so that transforming back to the ordinary frame of reference it has an energy $\varepsilon' = \gamma\bar{\varepsilon}(1+\cos\bar{\varphi}) = \gamma^2\varepsilon(1-\cos\vartheta)(1+\cos\bar{\varphi})$, and weighting the different angles ϑ of encounter, $\varepsilon' = \frac{4}{3}\gamma^2\varepsilon$ on average: on this basis X-rays of 30 keV to 1 MeV originate from electrons of 3 to 17 GeV. Felten (1965) showed that galactic electrons were insufficient to account for the intensity of the X-rays, but that electrons in intergalactic space formed a plausible source, where an electron spectrum $J(E)\,dE \propto E^{-3\cdot 6}\,dE$ would be necessary, the fact that this spectral index was greater by 1 than that of the normal electron spectrum being the natural consequence (as mentioned in section 7.5) of the energy losses, which vary as E^2 and operate over cosmological times. If this interpretation of the X-radiation is correct, it implies that (a) in the absence of this form of energy loss the electron intensity in intergalactic space would have been $\sim$ 30 times less than in the Galaxy, implying a very considerable injection of particles into space, and (b) that the spectrum of electrons at the Earth is definitely not from extragalactic sources.

Though it may turn out that emission of X-rays from distant galaxies makes an important contribution to this background flux, and that at lower energies thermal radiation from hot gas plays a part, one can see that further tools for studying cosmic radiation outside the Galaxy and at early epochs of the history of the universe may thus be offered.

Turning now to γ-rays, their detection could reveal the effects of π°-meson production in collisions of the nuclear component, and so could be very useful. Only very recently have attempts to detect γ-rays above 1 MeV produced firm positive results: after operating an experiment for many months in an Earth satellite, Clark, Garmire and Kraushaar (1968) have reported a flux of 100 MeV photons from the galactic centre and disc considerably above what was expected. Attempts should eventually succeed in detecting high-energy γ-rays from discrete objects by observing the directions of motion of the electron–positron pairs produced, and thus locating sources.

References

The publication date is followed by an asterisk* when it differs from the date of the work mentioned in the text. Papers reprinted in Part 2 are omitted from this list.

ALLEN, C. W. (1963) *Astrophysical Quantities*, 2nd ed. Univ. London, Athlone Press.

ALVAREZ, L. W. and COMPTON, A. H. (1933) *Phys. Rev*. **43,** 835.

ANAND, K. C., BISWAS, S., LAVAKARE, P. J., RAMADURAI, S. and SREENIVASAN, N. (1965) *Proc. IUPAP Conf. Cosmic Rays, London*, Vol. 1, p. 396.

ANAND, K. C., DANIEL, R. R. and STEPHENS, S. A. (1968) *Phys. Rev. Lett*. **20,** 764.

ANDERSON, C. D. (1932) *Science*, **76,** 238.

AUGER, P. and EHRENFEST, P., Jr. (1937) *J. de Physique*, **8,** 204.

AUGER, P., MAZE, R. and GRIVET-MAYER, T. (1938) *C.R. Acad. Sci. (Paris)*, **206,** 1721.

BAADE, W. (1956) *Bull. Astron. Inst. Netherl*. **12,** 312.

BALASUBRAHMANYAN, V. K., HAGGE, D. E., LUDWIG, G. H. and MCDONALD, F. B. (1965) *Proc. IUPAP Conf. Cosmic Rays, London*, Vol. 1, p. 427.

BALDWIN, J. E. (1954) *Nature*, **174,** 320.

BETHE, H. A. and HEITLER, W. (1934) *Proc. Roy. Soc*. **A146,** 83.

BHABHA, H. J. and HEITLER, W. (1937) *Proc. Roy. Soc*. **A159,** 432.

BLACKETT, P. M. S. and OCCHIALINI, G. P. S. (1933) *Proc. Roy. Soc*. **A139,** 699.

BLACKETT, P. M. S. (1938) *Nature*, **142,** 992.

BOTHE, W., GENTNER, W., MAIER-LEIBNITZ, H., MAURER, W., WILHELMY, E. and SCHMEISER, K. (1937) *Phys. Zeits*. **38,** 964.

BOWEN, S., MILLIKAN, R. A. and NEHER, H. V. (1938*) *Phys. Rev*. **53,** 855.

BRAY, A. D., CRAWFORD, D. F., JAUNCEY, D. L., MCCUSKER, C. B. A., POOLE, P. C., RATHGEBER, M. H., ULRICHS, J., WAND, R. H., WINN, M. M. and UEDA, A. (1964) *Nuovo Cimento*, **32,** 827.

BRODE (1950): per ROSSI, B. (1952) *High-energy Particles*, p. 177. Prentice-Hall.

BROOKE, G., HAYMAN, P. J., KAMIYA, Y. and WOLFENDALE, A. W. (1964) *Proc. Phys. Soc*. **83,** 853.

BURBIDGE, G. R. and HOYLE, F. (1964) *Proc. Phys. Soc*. **84,** 141.

CAMERON, A. G. W. (1959) *Astrophys. J*. **129,** 676.

CARMICHAEL, H. and DYMOND, E. G. (1939*) *Proc. Roy. Soc*. **A171,** 321.

CHAPMAN, S. and FERRARO, V. C. A. (1931) *Terr. Mag. Atm. Elect*. **36,** 77, 171.

CLARK, G. W., EARL, J., KRAUSHAAR, W. L., LINSLEY, J., ROSSI, B., SCHERB, F. and SCOTT, D. W. (1961) *Phys. Rev*. **122,** 637.

CLARK, G. W., GARMIRE, G. P. and KRAUSHAAR, W. L. (1968) *Astrophys. J.* **153,** L 203.

CLAY, J. (1927) *Proc. Roy. Acad. of Amsterdam*, **30,** 1115.

COCCONI, G., LOVERDO, A. and TONGIORGI, V. (1946) *Phys. Rev.* **70,** 852.

COCKE, W. J., DISNEY, M. J. and TAYLOR, D. J. (1969) *Nature*, **221,** 525.

COLGATE, S. A. and JOHNSON, M. H. (1960) *Phys. Rev. Lett.* **5,** 235.

COLGATE, S. A. and WHITE, R. H. (1963) *Astrophys. J.* **143,** 626. See previous ref.

COMPTON, A. H. (1932) *Phys. Rev.* **41,** 111, 681.

COWLING, T. G. (1957) *Magnetohydrodynamics.* Interscience, New York.

CRANSHAW, T. E. and GALBRAITH, W. (1954) *Phil. Mag.* **45,** 1109.

CRANSHAW, T. E. and GALBRAITH, W. (1957) *Phil. Mag.* **2,** 797, 804.

DAUDIN, J. (1945) *Ann. de Phys.* **20,** 563.

DICKE, R. H., PEEBLES, P. J. E., ROLL, P. G. and WILKINSON, D. T. (1965) *Astrophys. J.* **142,** 414. P.G.R. and D.T.W. *Phys. Rev. Lett.* **16,** 405.

DOBROTIN, N. A., DOVZENKO, O., ZATSEPIN, V., MURZINA, E., NIKOLSKII, S. I., RAKOBOLSKAYA, I. and TURKIS, E. (1958) Suppl. to *Nuovo Cimento*, **8,** 612.

DORMAN, L. I. (1963) *Progr. Cosmic Ray and Elementary Particle Physics*, Vol. 7. (Ed. J. G. Wilson and S. A. Wouthuysen.) North-Holland, Amsterdam.

DOVZENKO, O. I., ZATSEPIN, G. T., MURZINA, E. A., NIKOLSKII, S. I. and IAKOVLEV, V. I. (1960) *Proc. IUPAP Cosmic Ray Conf., Moscow*, Vol. 2, p. 134.

EARL, J. A. (1961) *Phys. Rev. Lett.* **6,** 125.

FELTEN, J. E. (1965) *Phys. Rev. Lett.* **15,** 1003.

FERMI, E. (1940) *Phys. Rev.* **57,** 485.

FERMI, E. (1954) *Astrophys. J.* **119,** 1.

FORBUSH, S. E. (1946) *Phys. Rev.* **70,** 771.

FORBUSH, S. E. (1954*) *J. Geophys. Res.* **59,** 525.

FREIER, P., LOFGREN, E. J., NEY, E. P., OPPENHEIMER, F., BRADT, H. L. and PETERS, B. (1948) *Phys. Rev.* **74,** 213.

FUJIMOTO, Y. and HAYAKAWA, S. (1967) *Handb. d. Phys. XLVI/2*, 115.

GINZBURG, V. L. (1953) *Dokl. Akad. Nauk. SSSR*, **92,** 727. (Transl. NSF-tr-207.)

GINZBURG, V. L. and SYROVATSKII, S. I. (1964) *The Origin of Cosmic Rays.* Pergamon, Oxford.

GLOECKLER, G. and JOKIPII, J. R. (1966) *Phys. Rev. Lett.* **17,** 203.

GOLD, T. (1968) *Nature*, **218,** 731; **221,** 25 (1969).

GORDON, I. M. (1954) *Dokl. Akad. Nauk. SSSR*, **94,** 413.

GREISEN, K. (1966) *Phys. Rev. Lett.* **16,** 748.

HARTMAN, R. C. (1967) *Astrophys. J.* **150,** 371.

HEITLER, W. and SAUTER, F. (1933) *Nature*, **132,** 892.

HESS, V. F. (1913) *Phys. Zeits.* **14,** 610.

HEWISH, A., BELL, S. J., PILKINGTON, J. D. H., SCOTT, P. F. and COLLINS, R. A. (1968) *Nature*, **217,** 709.

HILBERRY, N. (1941) *Phys. Rev.* **59,** 763; **60,** 1.

HOYLE, F. (1965) *Phys. Rev. Lett.* **15,** 131.

JANOSSY, L. and LOVELL, A. C. B. (1938) *Nature*, **142,** 716.

JANSKY, K. J. (1933*) *IRE Proc.* **21,** 1387; **23,** 1158 (1935).

JOHNSON, T. H. (1933) *Phys. Rev.* **43,** 834.

KIEPENHEUER, K. O. (1950) *Phys. Rev.* **79,** 738.

KLEIN, O. and NISHINA, Y. (1929) *Z. Phys.* **52**, 853.
KOLHÖRSTER, W. (1914) *Deutsch. Phys. Gesell. Verh.* **16**, 719.
KOLHÖRSTER, W., MATTHES, I. and WEBER, E. (1938) *Naturwiss.* **26**, 576.
KULENKAMPFF, H. (1938) *Deutsch. Phys. Gesell. Verh.* **19**, 92.
KULIKOV, G. V. and KHRISTIANSEN, G. B. (1958) *Zh. Eksper. Teor. Fiz.* **35**, 635. (Transl. *Soviet Phys. JETP* **8**, 441 (1959).)
KURIL'CHIK, V. N. (1965) *Astr. Zh.* **42**, 1138. (Transl. *Soviet Astr. AJ* **9**, 884.)
LANDAU, L. D. and RUMER, G. (1938) *Proc. Roy. Soc.* **A166**, 213.
LAPOINTE, M., KAMATA, K., GAEBLER, J., ESCOBAR, I., DOMINGO, V., SUGA, K., MURAKAMI, K., TOYODA, Y. and SHIBATA, S. (1968) *Can. J. Phys.* **46**, S 68.
LATTES, C. M. G., MUIRHEAD, H., OCCHIALINI, G. P. S. and POWELL, C. F. (1947) *Nature*, **159**, 694.
LEMAÎTRE, G. and VALLARTA, M. S. (1933) *Phys. Rev.* **43**, 87.
LINSLEY, J. and SCARSI, L. (1962) *Phys. Rev. Lett.* **9**, 123.
LINSLEY, J. (1963) *Proc. IUPAP Cosmic Ray Conf., Jaipur*, Vol. 4, p. 77.
LONGAIR, M. S. (1966) *Monthly Not. Roy. Astron. Soc.* **133**, 421.
MARSHAK, R. E. and BETHE, H. A. (1947) *Phys. Rev.* **72**, 506.
MATHIESEN, O., LONG, C. E., FREIER, P. S. and WADDINGTON, C. J. (1968) *Can. J. Phys.* **46**, S 583.
MCCRACKEN, K. G. (1962) *J. Geophys. Res.* **67**, 423, 435, 447.
MCDONALD, F. B. (1956) *Phys. Rev.* **104**, 1723.
MCDONALD, F. B. and WEBBER, W. R. (1962) *J. Geophys. Res.* **67**, 2119.
MILLER, J. S. and WAMPLER, E. J. (1969) *Nature*, **221**, 1037.
MILLIKAN, R. A. and CAMERON, G. H. (1926) *Phys. Rev.* **28**, 851.
MILLIKAN, R. A. and OTIS, R. M. (1926) *Phys. Rev.* **27**, 645.
NEHER, H. V., PETERSON, V. Z. and STERN, E. A. (1953*) *Phys. Rev.* **90**, 655.
NEHER, H. V. (1956) *Phys. Rev.* **103**, 228.
NEUGEBAUER, M. and SNYDER, C. W. (1962) *Science*, **138**, 1095.
NISHINA, Y., TAKEUCHI, M. and ICHIMIYA, T. (1937) *Phys. Rev.* **52**, 1198.
O'DELL, F. W., SHAPIRO, M. M. and STILLER, B. (1962) *J. Phys. Soc. Japan*, **17**, Suppl. A3, 23.
OKUDA, H. and TANAKA, Y. (1968) *Bull. Astron. Inst. Netherl.* **20**, 129.
OORT, J. H. and WALRAVEN, T. (1956) *Bull. Astron. Inst. Netherl.* **12**, 285.
OPPENHEIMER, J. R. and PLESSET, M. S. (1933) *Phys. Rev.* **44**, 53.
PAL, Y. (1967) *Handbook of Physics*, chapter 9. (Ed. E. U. Condon and H. Odishaw.) McGraw-Hill.
PARKER, E. N. (1957) *Phys. Rev.* **107**, 830.
PARKER, E. N. (1958a) *Phys. Rev.* **109**, 1328.
PARKER, E. N. (1958b) *Astrophys. J.* **128**, 664.
PARKER, E. N. (1963) *Interplanetary Dynamical Processes*. Wiley, New York.
PARKER, E. N. (1969) *Space Science Reviews*, **9**, 651.
PENZIAS, A. A. and WILSON, R. W. (1965) *Astrophys. J.* **142**, 19.
PERKINS, D. H. (1947) *Nature*, **159**, 126.
PETERS, B. (1958): *see* PAL, Y. (1967).
PETERS, B. (1960) *Proc. IUPAP Cosmic Ray Conf., Moscow*, Vol. 3, p. 157.
PFOTZER, G. (1936) *Z. Phys.* **102**, 23.

POWELL, C. F., FOWLER, P. H. and PERKINS, D. H. (1959) *The Study of Elementary Particles by the Photographic Method*. Pergamon, London.

PUPPI, G. (1956) *Progr. in Cosmic Ray Physics*, Vol. 3, p. 341. (Ed. J. G. Wilson.) North-Holland, Amsterdam.

QUENBY, J. J. and WENK, G. J. (1962) *Phil. Mag.* **81,** 1457.

QUENBY, J. J. (1967) *Handb. d. Phys. XLVI/2*, 310.

REBER, G. (1940) *IRE Proc.* **28,** 68. Also *Astrophys. J.* **100,** 279 (1944).

REGENER, E. (1931*) *Nature*, **127,** 233.

RICHTMYER, R. D. and TELLER, E. (1949) *Phys. Rev.* **75,** 1729.

ROCHESTER, G. D. and BUTLER, C. C. (1947) *Nature*, **160,** 855.

ROSE, D. C., FENTON, K. B., KATZMAN, J. and SIMPSON, J. A. (1956) *Can. J. Phys.* **34,** 968.

ROSSI, B. (1930) *Nature*, **125,** 636.

ROTHWELL, P. and QUENBY, J. J. (1958) Suppl. to *Nuovo Cimento* **8,** 249.

RUTHERFORD, E. and COOKE, H. L. (1903) *Phys. Rev.* **16,** 183.

RUTHERFORD, E. (1913) *Radioactive Substances and Their Radiations*, p. 596. Cambridge U.P.

SAKAKIBARA, S. (1965) *Proc. IUPAP Conf. Cosmic Rays, London*, Vol. 1, p. 135.

SAKATA, S. and INOUE, T. (1946) *Progr. Theor. Phys.* **1,** 143.

SCHWINGER, J. (1949) *Phys. Rev.* **75,** 1912.

SHEA, M. A. and SMART, D. F. (1967) *J. Geophys. Res.* **72,** 2021.

SHKLOVSKII, I. S. (1960) *Cosmic Radio Waves*. Harvard U.P.

SHKLOVSKII, I. S. (1966) *Astr. Zh.* **43,** 10. (Transl. *Soviet Astr. AJ* **10,** 6.)

SKOBELZYN, D. (1929) *Z. Phys.* **54,** 686.

SKOBELZYN, D. V., ZATSEPIN, G. T. and MILLER, V. V. (1947) *Phys. Rev.* **71,** 315.

SNYDER, H. S. (1938) *Phys. Rev.* **53,** 960; **76,** 1563 (1949).

STERNHEIMER, R. M. (1961) In *Methods of Experimental Physics*, Vol. 5A. (Ed. L. C. L. Yuan and C. S. Wu.) Academic Press, New York.

STREET, J. C. and STEVENSON, E. C. (1937) *Phys. Rev.* **52,** 1003.

SUESS, H. and UREY, H. C. (1956) *Rev. Mod. Phys.* **28,** 53.

SWANN, W. F. G. (1938) *J. Frankl. Inst.* **226,** 598.

SWANN, W. F. G. (1941) *Phys. Rev.* **59,** 770.

SYROVATSKII, S. I. (1961) *Zh. Eksper. Teor. Fiz.* **40,** 1788. (Transl. *Soviet Phys. JETP* **13,** 1257.)

TER HAAR, D. (1950) *Rev. Mod. Phys.* **22,** 119.

UNSÖLD, A. (1951) *Phys. Rev.* **82,** 857.

VAN ALLEN, J. A. and SINGER, S. F. (1950) *Phys. Rev.* **78,** 819; Correction, **80,** 116.

VON WEISZÄCKER, C. F. (1934) *Z. Phys.* **88,** 612.

WADDINGTON, C. J. (1956) *Nuovo Cimento*, **3,** 930.

WEBBER, W. R. (1967) *Handb. d. Phys. XLVI/2*, 181.

WEBBER, W. R. and ORMES, J. F. (1967) *J. Geophys. Res.* **72,** 5957.

WILLIAMS, E. J. (1934) *Phys. Rev.* **45,** 729.

WILSON, C. T. R. (1901) *Proc. Roy. Soc.* **68,** 151.

WINCKLER, J. R., STIX, T., DWIGHT, K. and SABIN, R. (1950) *Phys. Rev.* **79,** 656.

YIOU, F., DUFAURE DE CITRES, J., FREHEL, F., GRADSTAJN, E. and BERNAS, R. (1968) *Can. J. Phys.* **46,** S 557.

YUKAWA, H. (1935) *Proc. Phys.-Math. Soc. Japan*, **17,** 48.

PART 2

1. Observations of the Penetrating Radiation on Seven Balloon Flights†

VIKTOR F. HESS‡

LAST year I had the opportunity of making two balloon ascents to investigate the penetrating radiation: the first flight has already been reported at the scientific meeting at Karlsruhe.[1] On both journeys no essential change was found in the radiation up to an altitude of 1100 m, compared with that observed near the ground. Gockel[2] had similarly been unable to find the expected decrease in the radiation with height in two balloon ascents. The conclusion was thence drawn that there must be another source of the penetrating radiation in addition to the γ-radiation from the radioactive substances in the Earth's crust.

A grant from the Kaiserlichen Akademie der Wissenschaften in Vienna has enabled me to carry out a series of seven more balloon ascents this year, from which was obtained more comprehensive observational material, extended in many ways.

Two Wulf radiation detectors of 3 mm wall thickness served for the observation of the penetrating radiation in the first place, perfectly air-tight and capable of withstanding the pressure variations on all the ascents. Instrument 1 had an ionization volume of 2039 cc, the capacity being 1·597 cm; instrument 2 had volume 2970, the capacity amounting

† *Physik. Zeitschr.* **13,** 1804–1091 (1912). (From the Section on Geophysics, Meteorology and Geomagnetism.)

‡ Vienna.

[1] *Physik. Zeitschr.* **12,** 998–1001 (1911); *Wien. Sitz.-Ber.* **120,** 1575–1585 (1911).

[2] *Physik. Zeitschr.* **12,** 595–597 (1911).

to 1·097 cm. Hence a loss of charge of 1 volt per hour corresponded in instrument 1 to an ionization rate of $q = 1{\cdot}56$ ions per cc per sec, and in 2 to $q = 0{\cdot}7355$ ions per cc per sec.

Both instruments had been electrolytically galvanized on the inside to reduce the radiation from the walls of the vessel. This was suggested by Dr. Bergwitz. After this treatment detector 1 indicated a normal ionization of about 16 ions, detector 2 of 11 ions per cc per sec. The firm of Günther and Tegetmeyer in Brunswick has also provided another essential improvement to the apparatus: hitherto the sharp setting on the fibres took place solely by moving the eyepiece, which gave rise to a not unimportant change in the magnification, and by repeated adjustment produced differences of up to 0·5 in the readings. This firm has now mounted a sliding negative lens in the eyepiece tube, which permits focusing for various positions of the fibres without a perceptible change of magnification resulting. The accuracy of setting is thus considerably increased.

The wall thickness of the instruments 1 and 2 was 3 mm, so that essentially only the γ-rays could be effective. In order to study simultaneously the behaviour of β-rays I also used a third instrument, which was not made air-tight, but consisted of an ordinary Wulf two-fibre electrometer over which was inverted a cylindrical ionization vessel of 16·7 litre volume made of the thinnest commercially available zinc (wall thickness 0·188 mm), so soft rays with the character of β-rays could also play an effective part. A zinc spike 20 cm high fixed to the fibre support of the electrometer acted as charge disperser. The capacity was 6·57 cm.

The insulation loss in the thick-walled Wulf radiation detectors 1 and 2 was determined by the usual way with a lowered guard tube. The hourly loss of charge amounted to 0·2 volt in instrument 1, and 0·7 volt in no. 2. A breakdown of insulation due to damp weather is never found.

* * * * *[a]

[a] Superscript letters refer to Notes at end of articles.

The last and most important point of the investigation was the measurement of the radiation at the greatest possible heights. Whilst in the six ascents undertaken from Vienna the weak lifting power of the local gas, as well as the meteorological conditions, had not permitted this, in an ascent made with hydrogen from Aussig on the Elbe I managed to carry out measurements up to 5350 m altitude.

For several hours before each flight control observations with all three detectors were made. For this the instruments were attached by means of brackets to the balloon basket, exactly as during the flight itself. The observations before the ascent were carried out at the club site of the k. k. Österreichischen Aeroklub, a level grass plot in Prater in Vienna. L. V. King[3] has briefly conjectured that balloon observations could be disturbed by the proximity of weakly radioactive ballast sand. An increase in the radiation was never found from supplies of ballast sand.

In instruments 1 and 2 the same air density always prevails inside the ionization space as at the site of the ascent (on average 750 mm). On the other hand in the thin-walled detector 3, the pressure is always the same as in the surroundings. A reduction of the directly observed data is therefore necessary, especially in observations at great heights.

* * * * *

In the following tables q_1, q_2, q_3 denote the penetrating radiation observed with the three instruments 1, 2, 3, in ions per cc per sec. The elementary charge is taken as $e = 4{\cdot}65 \times 10^{-10}$ esu. The mean height of the balloon during a particular observation period (usually 1 hour) was deduced from the barograph trace by a graphical method. A mean value for the relative height was then calculated from the altitude above sea level of the locality directly below. The hour of day is given by the 24-hour clock in the tables.

A full account of all the balloon observations has been presented to the Kaiserlichen Akademie der Wissenschaften in Vienna, and published in the proceedings. Here I will give detailed accounts of only the two most important ascents and be content to quote average values for the others.

[3] *Phil. Mag.* **23** (6), 242 (1912).

1st Flight

This took place on the occasion of the very considerable partial eclipse of the Sun in Lower Austria on 17th April 1912. It was observed from 11 a.m. to 1 p.m. at an absolute altitude of 1900–2750 m above an almost complete layer of cumulus. No reduction in the penetrating radiation was observed during the advancing eclipse. Instrument 2 showed, e.g., an ionization of 10·7 ions before the ascent, at an average relative height of 1700 m, 11·1 ions, later at 1700 to 2100 m, during the first phase of the eclipse 14·4, and then at about 50% solar obscuration 15·1 ions. Further measurements were not possible, since the balloon was forced to descend as a result of the cooling of the gas.

An increase in the radiation was thus found at around 2000 m. As no effect of the eclipse was to be seen on the penetrating radiation, we may conclude that if a part of the radiation is of cosmic origin it can hardly come from the Sun, at least so long as one thinks of a γ-radiation propagated in straight lines. That in subsequent balloon flights I never found a distinct difference in the radiation between day and night confirms this view.

* * * * *[a]

The following 7th flight was undertaken as a really high ascent.

7th Flight (7th August 1912)

We took off at 6.12 a.m. from Aussig on the Elbe. We flew over the Saxony border by Peterswalde, Struppen near Pirna, Bischofswerda and Kottbus. The height of 5350 m was reached in the region of Schwielochsee. At 12.15 p.m. we landed near Pieskow, 50 km east of Berlin. Unfortunately no observations could be made at the site of the ascent before the journey, but measurements were made after landing under the still filled balloon, in order to see immediately after descending from 5000 m whether the balloon had become covered with radioactive induction and so emitted a radiation itself. As may be seen from the table (Obs. no. 11), no sign of an increase in radiation was observed

Balloon "Böhmen" (1680 cbm hydrogen) Leader: Captain W. Hoffory.
Meteorological observer: E. Wolf. Electr. observer: V. F. Hess.

No.	Time	Mean height abs. m	Mean height rel. m	Observed radiation Inst. 1 q_1	Inst. 2 q_2	Inst. 3 q_3	Inst. 3 red. q_3	Temp.	Rel. humidity %
1	15^h15–16^h15	156	0	17·3	12·9	—	—	1½ days before the ascent (in Vienna)	
2	16^h15–17^h15	156	0	15·9	11·0	18·4	18·4		
3	17^h15–18^h15	156	0	15·8	11·2	17·5	17·5		
4	6^h45– 7^h45	1700	1400	15·8	14·4	21·1	25·3	+6·4°	60
5	7^h45– 8^h45	2750	2500	17·3	12·3	22·5	31·2	+1·4°	41
6	8^h45– 9^h45	3850	3600	19·8	16·5	21·8	35·2	−6·8°	64
7	9^h45–10^h45	4800 (4400–5350)	4700	40·7	31·8	(ended by accident)		−9·8°	40
8	10^h45–11^h15	4400	4200	28·1	22·7			—	—
9	11^h15–11^h45	1300	1200	(9·7)	11·5			—	—
10	11^h45–12^h10	250	150	11·9	10·7			+16·0°	68
11	12^h25–13^h12	140	0	15·0	11·6	(after landing at Pieskow, Brandenburg)			

under the balloon after landing. The weather was not perfectly clear on this journey: a barometric depression approaching from the west made itself apparent as clouding set in. It should nevertheless be expressly state that we were never in a cloud nor even in the vicinity of one; as at the time that the cumulus clouds appeared as isolated balls scattered all around the horizon, we were already above 4000 m in height. Above us as we approached the maximum height was a thin cloud sheet much higher still, whose base was probably at 6000 m at least, and through which the Sun shone only weakly.

We first consider the results with the thick-walled instruments 1 and 2. At 1500 to 2500 m mean altitude the radiation was just about as strong as was normally found on the ground. *Then begins a clearly perceptible rise in the radiation with increasing height, seen by both instruments*—at 3600 above the ground both values are already about 4–5 ions higher than on the ground.

* * * * *

In both γ-ray detectors the values at the greatest altitude are about 20 to 24 ions higher than at the ground. The very high values $q_1 = 28 \cdot 1$ and $q_2 = 22 \cdot 7$ were found also on the descent at 4400 m. These greatly exceed the normal values of 12 and 11 ions. In the ensuing very rapid descent (2 m per sec) the very low value of 9·7 was measured in instrument 1 at a mean height of 1200 m, while instrument 2 registered the normal value 11·5. I think it possible that in instrument 1, which was provided with very thick fibres, a certain stiffness of the filament sometimes produced a noticeable irregularity.

The results obtained from both detectors under the still-inflated balloon after landing were, as remarked above, quite normal.

To gain a picture of the variation of the penetrating radiation with altitude, on the average, I have combined all the 88 radiation values I have observed in balloons at suitable intervals of height in the following table. For each height, averages are formed from several individual values which were obtained under different conditions and which could be influenced by the previously mentioned transient variations, so one should not expect to obtain as yet a wholly exact picture of the development of the radiation with increasing altitude. The numbers in

TABLE OF MEAN VALUES

Mean height above ground m	Observed radiation in ions per cc per sec			
	Inst. 1	Inst. 2	Inst. 3	
	q_1	q_2	q_3 red.	q_3 not red.
0	16·3 (18)	11·8 (20)	19·6 (9)	19·7 (9)
up to 200	15·4 (13)	11·1 (12)	19·1 (8)	18·5 (8)
200– 500	15·5 (6)	10·4 (6)	18·8 (5)	17·7 (5)
500–1000	15·6 (3)	10·3 (4)	20·8 (2)	18·5 (2)
1000–2000	15·9 (7)	12·1 (8)	22·2 (4)	18·7 (4)
2000–3000	17·3 (1)	13·3 (1)	31·2 (1)	22·5 (1)
3000–4000	19·8 (1)	16·5 (1)	35·2 (1)	21·8 (1)
4000–5200	34·4 (2)	27·2 (2)	—	—

brackets indicate the numbers of actual observations from which the means were derived.

We learn from this table that directly above the earth the radiation falls off a little. On the average this decrease amounts to 0·8 to 1·4 ions. Since, however, a decrease of up to 3 ions has been found on some ascents, and over 2 ions in very many measurements, we shall quote about 3 ions as the maximum value of the decrease. This decrease extends up to 1000 m from the ground. It clearly results, as mentioned earlier, from the absorption of γ-rays emitted from the earth's surface. Hence we conclude that *the γ-radiation from the earth's surface and the uppermost layers of the ground gives an ionization of about 3 ions per cc per sec in zinc vessels.*

At altitudes up to 2000 m the radiation again begins to increase perceptibly. By 3000 to 4000 m the increase amounts to 4 ions, and at 4000 to 5200 m fully 16 to 18 ions, in both detectors. With the thin-walled detector, when the values are reduced to normal pressure, the decrease is reversed sooner and more strongly.

What is the cause of this increase of the penetrating radiation with altitude, now observed many times and simultaneously with three detectors?

If one adopts the view that only the known radioactive substances in the earth's crust and atmosphere emit a radiation with the character of γ-radiation and produce ionization in closed vessels, serious difficulties defy explanation:

According to the direct determinations of the absorption coefficient of γ-rays in air by myself[4] and Chadwick,[5] the absorption of the radiation from the earth's surface must be rapid, so that 500 m above the ground scarcely 10% of the radiation can remain. As has been mentioned, I have been able to demonstrate this experimentally in the balloon; but at the same time it appears that the radioactive substances in the ground do not play such a preponderant role in the total radiation as many authors believe. Its share was determined as 3 ions per cc per sec.

[4] *Wien. Sitz.-Ber.* **120,** 1205–1212 (1911).
[5] *Le Radium,* **9,** 200–202 (1912).

There still remain the decay products of emanations, for producing ionization through γ-radiation at high altitudes. Because of their short lifetimes, thorium and actinium emanations and their decay products will not be able to reach great heights. Only radium-emanation, with a half-life of nearly 4 days, can be drawn up to great altitudes in rising air currents. In general, however, the concentration of emanation, and therefore also the RaC content of the air, will soon decrease with height. An increase in the radiation with height could only arise through an accidental accumulation of RaC of purely local character: it is for instance imaginable that such accumulations might occur in stable layers at a temperature inversion, or in cumulus clouds or in mist, as it is known that RaC atoms frequently act as condensation nuclei. However, a uniform increase of the penetrating radiation with altitude, as found in my observations, cannot be explained in this way. Also, in flights 2 and 6, in which the balloon travelled close to the ground for hours in such a stable layer at an inversion, I observed no increase in the radiation, although the RaC content of the air must be larger near the ground. At a height of 5000 m the RaC content will certainly not suffice to cause so great an increase in the radiation as I found.

The fluctuations in the radiation often found by Pacini[6] and Gockel[7] on the sea and on land, and by myself in the balloon, also raise great difficulties for an explanation of the penetrating radiation based only on the radioactive theory. I have repeatedly observed such variations in the middle of the night, in a perfectly calm atmosphere. In the absence of any meteorological change, there are no grounds for tracing them back to changes in the distribution of radioactive substances in the atmosphere.

The results of the present observations seem to be most readily explained by the assumption that a radiation of very high penetrating power enters our atmosphere from above, and still produces in the lowest layers a part of the ionization observed in closed vessels. The intensity of this radiation appears to be subject to transient variations, recognizable in hourly readings. Since I found a reduction in the radiation at the bal-

[6] *Le Radium*, **8,** 307–312 (1911).
[7] *Jahrb. d. Rad. u. Elektron.* **9,** 1–15 (1912).

loon neither by night nor at a solar eclipse, one can hardly consider the Sun as the origin of this hypothetical radiation, at least so long as one thinks only of a direct γ-radiation with rectilinear propagation.

It is not so very surprising that the increase in the radiation first becomes really noticeable beyond 3000 m: the decrease in the γ-radiation from the ground prevails in the first 1000 m, and after that comes the decrease in the induction strength, which makes itself felt until over 3000 m. The absorption of the radiation coming from above in any case follows an exponential curve; the increase in radiation as one goes up will hence become steeper at greater heights.

Note

[a] Hess was concerned amongst other things with the apparent fluctuations in the radiation, but we omit this material.

2. The Nature of the High-altitude Radiation†

W. BOTHE and W. KOLHÖRSTER

When two Geiger–Müller counter tubes are placed close together it is found that amongst the deflections caused by the high-altitude radiation, a considerable fraction occur simultaneously in both counters. From the dependence of these coincidences on the relative positions of the counters, and from their great frequency, it is concluded that they signify the passage of single corpuscular rays through both counting tubes. The absorbability of these corpuscular rays was determined by measuring the diminution of the coincidences when absorbing sheets were placed between the two counters. This showed that the corpuscular radiation is just as strongly absorbed as the high-altitude radiation itself. Hence it is concluded that the high-altitude radiation, as far as it manifests itself in the phenomena so far observed, is of a corpuscular nature. Its probable properties are discussed from this viewpoint.

I. Object and Method of the Investigation

1. OBJECT OF THE EXPERIMENT

Research into the high-altitude radiation has so far taken a strange course, for the most diverse features of the radiation, such as intensity, distribution, absorption and scattering, and even its origin, are investigated and debated, whilst the really essential question regarding the nature of the high-altitude radiation has hitherto found no experimental answer. The main reason for this is that owing to the low intensity and great penetrating power of these rays it is not possible directly to examine a screened beam of rays. If the view that the high-altitude radiation is a very hard γ-radiation has until now been universally pre-

† *Zeitschrift für Physik*, **56**, 751–777 (1929).

ferred, this has only been because of the enormous penetrating power, which would be more difficult to explain by a corpuscular radiation. Still, nearly all the authors concerned have referred to the possibility in principle of an electron radiation. In the rather different context of the influx theory for the maintainance of the Earth's negative charge, Schweidler, Swann and Hoffmann, following an idea of Simpson,[1] looked for a penetrating electron radiation directly, investigating whether an exposed insulated metal block gradually charged itself up: the result of these experiments was negative. On the other hand, in his fine Wilson photographs, Skobelzyn[2] found electron tracks which were not of radioactive origin, whose great magnetic rigidity indicated an energy of at least $1{\cdot}5 \times 10^7$ eV. One must assume with Skobelzyn that these fast electrons are connected with the high-altitude radiation; for it can scarcely be doubted that the ionization through which the high-altitude radiation is normally measured has its immediate origin in a corpuscular radiation. The essential problem is thus the following: Is this corpuscular radiation to be understood as secondary to a γ-radiation, as has been customary, or does it represent in itself the high-altitude radiation?

To answer this question it is especially important to know the penetrating power of the corpuscular radiation. If this corresponds to that of the high-altitude radiation itself, it will strongly support the hypothesis that the high-altitude radiation itself has a corpuscular nature. If on the other hand the corpuscular radiation is markedly softer than the high-altitude radiation, the latter must be a γ-radiation which produces the corpuscular radiation by impact on matter. We have now performed an experiment to determine the penetrating power of the supposed secondary electrons, proceeding from the above γ-ray hypothesis. It turned out that this experiment made it possible to decide between the above-mentioned interpretations.

[1] G. C. Simpson, *Nature*, **69**, 270 (1904); E. v. Schweidler, *Wien. Ber.* **127**, 515 (1918); *Ann. d. Phys.* **63**, 726 (1920); W. F. G. Swann, *Terr. Magn. and Atm.* **20**, 105 (1915); *Phys. Rev.* **9**, 555 (1917); *J. Frankl. Inst.* **203**, 11 (1927); G. Hoffmann, *Phys. Zs.* **27**, 296 (1926).

[2] D. Skobelzyn, *Zs. f. Phys.* **43**, 371 (1927); **54**, 686 (1929).

2. EXPERIMENTAL METHOD

In the domain of radioactive γ-rays the absorption coefficients of secondary electrons have often been measured by letting an electron-free beam of γ-rays fall on varying layers of the substance under test, and measuring the rise of the resulting electron radiation with increasing layer thicknesses. This method is not applicable to a high-altitude γ-radiation, however, since this has had ample opportunity to produce secondary electrons during its passage through the atmosphere, and it is not feasible to obtain a beam of high-altitude radiation freed from these penetrating electrons.

* * * * *

Hence one is forced to set out on the pursuit of individual radiation particles by a counter method.[3] Geiger's point counter was hardly suitable for such investigations, as it counts over too small a volume. Geiger and Müller's[4] recently perfected type of counter, the "electron counter tube", has for the first time opened up the prospect of a successful treatment of these problems, for to this instrument can easily be given any sensitivity which may be necessary as it can be extended to increase the effective volume almost without limit.

Since as soon as two counters were placed together coinciding responses were seen, the use of the coincidence principle suggested itself in the following form for the identification of an individual radiation particle. The responses of both counters were recorded as usual on a common film, whereby the passage of the same radiation particle through both counters showed up sharp and clear as a coincidence between the deflections on both sides. If absorbing sheets were then placed between the two counters, the frequency of the coincidences could be recorded as a function of absorber thickness, to display directly the absorption curve of the corpuscular radiation.

This was the basic idea of the experimental method, the correctness

[3] V. F. Hess and R. W. Lawson have previously used an electrical counting method on the high-altitude radiation (*Wien. Ber.* **125**, 285, 1916).

[4] H. Geiger and W. Müller, *Die Naturwiss.* **16**, 617, 1928. The importance of this instrument for research on high-altitude radiation was emphasized there.

of which will be more thoroughly verified below by detailed experimental and theoretical discussion. It is obvious that this method calls for a very extensive series of experiments if it is to yield reliable results; the maximum frequency of coincidences in our main experiment amounted to only 2·7/min. In fact about 90,000 individual deflections were registered for the moderate accuracy obtained here. In spite of this drawback, however, the method seems to us irreplaceable at present.

II. Main Experiment

3. ARRANGEMENT AND METHOD OF EVALUATION

In Figure 1 is shown the arrangement which, with a few modifications, was used for all experiments on the high-altitude radiation. Both the counters Z_1, Z_2, had inner diameters 5 cm, and length 10 cm; they were originally made of brass 1 mm thick, but later of 1 mm thick zinc, and were closed at the ends by ebonite plugs which supported the central wires. They were prepared by the method of Geiger and Müller. The

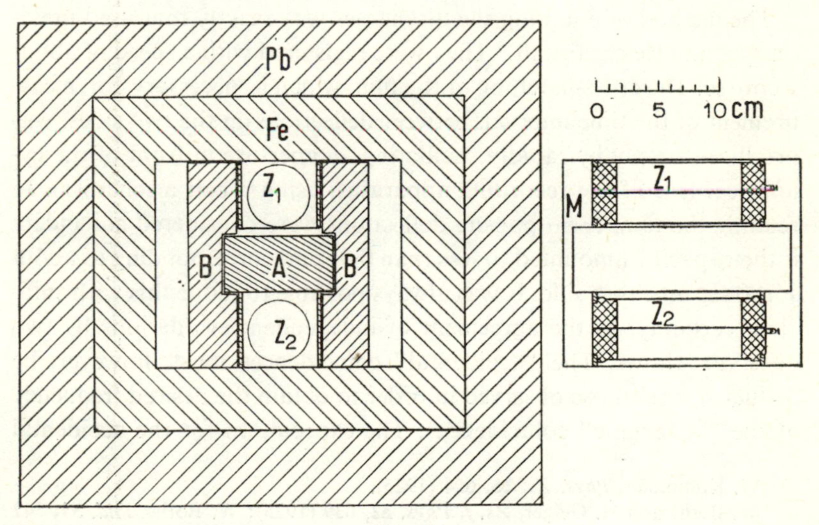

FIG. 1.

counters were filled to 4 to 6 cm Hg pressure with dry air free from carbon dioxide and emanation. They were supported by a brass frame *M*, which was so arranged that absorbing sheets up to 45 mm thick could be carried between the counters. At the sides the tubes were shielded by lead blocks *BB*; these had slots in which the absorber fitted. The thickness of these side screens was such that a radiation particle emerging from the counter which might by chance be scattered round the absorber into the other counter had to penetrate a significantly greater mass than the absorber offered. The whole apparatus was contained in a shield of 5 cm iron (inside) plus 6 cm lead (outside). The lead was about 150 years old and found to be free from γ-radiation. This shielding sufficed to make the radiation from external radioactivity negligible in practice. Insulated leads were conveyed through the shield both for the high tension and the connection between the central electrodes and the electrometers, which were the "filament electrometers" described by one of us,[5] used in the "needle connection". The voltage for each counter was carefully adjusted for optimum working, coming to about -1300 V, and was constant for a given counter over the whole duration of the experiment (3 months so far).

The method of counting the discharges was exactly that used previously at the Reichsanstalt.[6] They were registered with a speed of about 1 cm/sec. The cross-hatching of the film, which is necessary for measurement of the time intervals between deflections on the two sides, was produced simply by rapidly oscillating an aperture in front of the cylindrical lens of the recording apparatus, using a bell mechanism. In reading the film, two opposite deflections were considered coincident if their spacing amounted to less than 0·01 cm, corresponding to about 1/100 sec on either side. It was always possible to judge this with sufficient certainty, as the agreement of independent evaluations by two observers shows. The "accidental" coincidences must, however, be deducted from those observed in order to obtain the desired frequency of the "systematic" coincidences. The expectation for the accidental

[5] W. Kolhörster, *Phys. Zs.* **26,** 654 (1925).

[6] W. Bothe and H. Geiger, *Zs. f. Phys.* **32,** 639 (1925); W. Bothe, *ibid.* **37,** 547 (1926).

coincidences on a length l cm of film amounts to $2\times0{\cdot}01\times N_1N_2/l$, if N_1, N_2 denote the respective numbers of deflections in this length.

There remained the possibility that the deflections began after a variable time-lag, as had previously been established for the point counter.[7] There would then have been the risk of missing real coincidences, because the corresponding deflections fell more than 0·01 cm apart. That this was not the case could be seen from the frequency of intervals between 0·01 and 0·02 cm, which agreed with the above expression, which also gives the expectation for this.

Table 1 serves as an example of a series of experiments and their evaluation.

The experiments were carried out in two places, part on the ground floor, and part in the attic of the main building of the Reichsanstalt. In the lower room, the high-altitude radiation had to pass through a concrete covering of about 2 m water-equivalent altogether, and then the metal shield of 1 m water-equivalent, before it entered the experimental apparatus; but only a thin skylight in the upper room, as the shield was removed.

When the counter arrangement was put in the shield, the frequency of deflections of both instruments fell considerably. The remaining effect, which can no longer be accounted for by the radioactivity of the surroundings, is at least to a large extent to be attributed to the high-altitude radiation, as Geiger and Müller have shown.[8] Also the frequency of these deflections agrees on the whole with the data of Geiger and Müller, when converted to the same counting area.

* * * * *

From the records of these deflections there appeared immediately so large a number of coincidences as could only be explained by corpuscular rays (see the full discussion in section 9).

[7] W. Bothe and H. Geiger, *loc. cit.*
[8] H. Geiger and W. Müller, *Phys. Zs.* **29**, 839 (1928).

4. EXPERIMENTS WITH FILTERED HIGH-ALTITUDE RADIATION

The first experiments were performed on the ground floor of the main building of the P.T.R. It soon stood out that the radiation which caused the coincidences must be extraordinarily penetrating, as in a short series of experiments the insertion of 1 cm and even 4 cm of lead as absorber gave no distinct decrease in the coincidences. It did not seem advisable to use still thicker lead sheets as, in addition to other inconveniences, the whole apparatus including the shield would then have had to be made larger. So for the final experiments the absorber was a block of pure gold (995·6), which was placed at our disposal for a short time. The block had a mean thickness of 4·1 cm and an area of 8·9×17·5 cm.

A longer series of experiments was performed with the lead block, again on the ground floor. A single experiment, of which Table 1 gives

TABLE 1. GROUND FLOOR. FILM 14

Recording time min	Absorber cm Au	Film length cm	Deflections above	Deflections below	Coincidences (counted)	Random coincs. (calc.)	Systemat. coincs. (calc.)
7	0	391	310	176	10	2·8	7·2
15	4·1	894	623	323	36	4·5	31·5
15	0	823	619	364	33	5·5	27·5
15	4·1	878	625	324	27	4·6	22·4
8	0	496	339	187	13	2·6	10·4

an example, comprised as a rule a 60-minute counting period, during which time 15 minutes with and without the gold block were recorded alternately. The experiments are summarized in Table 2. All the results show that the introduction of the gold block influences neither the number of deflections nor that of coincidences by significantly more than the statistical errors.[9] If, from the relative difference of the num-

[9] The mean error (m.d.) of the difference of two normally fluctuating counts $a-b$ is $\sqrt{(a+b)}$, the mean square errors a and b being added.

bers of coincidences with and without the gold, one calculates the purely formal mass-absorption coefficient (μ/ϱ) in gold, without at first discussing the meaning of this quantity, one finds $\mu/\varrho = (0{\cdot}2 \pm 0{\cdot}9)10^{-3}$ cm²/g.

TABLE 2. GROUND FLOOR

Film	Without absorber				With 4·1 cm gold			
	Recording time min	Deflections above	Deflections below	System. coincs.	Recording time	Deflections above	Deflections below	System. coincs.
9	16	633	428	23·3	16	642	429	33·8
10	20	726	468	29·4	20	742	460	34·3
11	26	959	642	50·0	30	1130	721	59·8
12	32	1252	766	60·0	30	1109	666	57·3
13	30	1168	668	38·1	30	1156	640	35·6
14	30	1268	727	45·1	30	1248	647	53·9
15	27	1249	625	52·1	30	1315	662	43·5
16	30	1256	661	54·8	30	1263	615	38·1
sum:	211	8511	4985	352·8	216	8605	4840	356·3
in 216 min	—	8713	5003	361·2	—	8605	4840	356·3
per min	—	40·3	23·6	1·67	—	39·8	22·4	1·65

Decrease in coincidences in 216 min (±m.d.): 5·4±26·8 = (1·5±7·4)%.

This result already allowed us to suppose that the "atmospheric" corpuscular radiation which generates the coincidences is not to be interpreted as β-radiation secondary to a high-altitude γ-radiation. For the mass-absorption coefficient of the weakly-filtered high-altitude radiation itself is of the order of magnitude 10^{-3}, so all the expectation is that the secondary radiation which it generates must be considerably softer. The further surmise, that the decrease in frequency of the coincidences follows the absorption curve of the high-altitude radiation itself, then suggested itself.

* * * * *

5. EXPERIMENTS WITH UNFILTERED HIGH-ALTITUDE RADIATION

The experiment was then set up in the attic of the same building where, following the removal of the shielding, the whole of the high-altitude radiation could act directly on the counters. The side walls of the shield projected 9 cm above the upper edge of the upper counter. Otherwise the whole arrangement was the same as before, and the experiments were carried out exactly as before. Table 3 shows the results. It confirmed our expectation: not only the numbers of deflections but also the coincidences are greater than on the ground floor, correspond-

TABLE 3. UPPER FLOOR

Film	Without absorber				With 4·1 cm gold			
	Recording time min	Deflections above	Deflections below	System. coincs.	Recording time min	Deflections above	Deflections below	System. coincs.
17	30	1 819	1 007	85·7	30	1 862	930	70·6
18	30	1 604	1 092	84·5	30	1 702	998	81·7
19	30	1 621	1 009	81·2	30	1 587	912	55·3
20	30	1 589	1 043	95·3	30	1 487	832	59·8
21	25	1 331	808	67·6	30	1 570	843	55·3
22	30	1 556	894	70·3	30	1 564	861	57·3
23	28	1 540	1 009	87·2	30	1 630	880	60·2
24	29	1 642	1 038	71·7	30	1 716	881	66·1
25	30	1 688	1 019	88·0	30	1 720	885	65·2
26	30	1 669	1 025	81·9	30	1 598	840	49·4
27	29	1 590	934	76·3	30	1 702	801	61·3
28	37	2 137	1 263	90·8	30	1 676	899	61·0
sum:	358	19 786	12 141	980·5	360	19 814	10 562	743·2
in 360 min	—	19 897	12 209	986·0	—	19 814	10 562	743·2
per min	—	55·3	33·9	2·74	—	55·0	29·3	2·06

Decrease in coincidences in 360 min: 242·8±41·6 = (24·6±4·2)%.

ing to the greater intensity of the high-altitude radiation. On insertion of the gold block the deflections of the lower counter and above all the coincidences clearly decrease, while the deflections of the upper counter are unchanged. From the coincidences, the mass-absorption coefficient in the previously mentioned formal sense works out to be

$$\frac{\mu}{\varrho} = (3{\cdot}5 \pm 0{\cdot}5) \times 10^{-3}\ \mathrm{cm}^2/\mathrm{g}.$$

In various substances, mass-absorption coefficients for the unfiltered high-altitude radiation between 2 and 5×10^{-3}, and occasionally higher still, have been found. Hence as the coincidences decrease at least approximately according to the absorption curve of the high-altitude radiation itself, this clearly suggests that the high-altitude radiation only manifests itself as a corpuscular radiation. The justification for this supposition can best be demonstrated through the following experimental and theoretical discussion.

* * * * *[a]

IV. Discussion

9. NATURE OF THE RAYS PRODUCING COINCIDENCES

From the magnitude of the coincidence fraction (the ratio of coincidences to deflections in the lower counter) one can in principle calculate the probability with which a single ray actuates one of the two counters: the fraction is proportional to this probability. One can then reach further conclusions as to the nature of the rays which produce the coincidences. Strictly, however, one must also know the directional distribution of the rays. As this was not sufficiently well known, the calculation was carried out under the simplest assumption, that the radiation is isotropic.

Let $F(\vartheta, \varphi)$ be the area of one counter when projected in the direction given by the polar angles ϑ, φ, and $f(\vartheta, \varphi)$ the part of this area

[a] Superscript letters refer to Notes at end of article.

which it has in common with the projection of the other counter (Figure 2). If each counter detects each ray passing through it, the coincidence fraction should be given by

$$r = \frac{\int f.d\Omega}{\int F.d\Omega},$$

where from symmetry the integral need only be taken over an octant of a sphere. The detailed evaluation of this expression was carried out (partly graphically) for the tube separation of 10·4 cm as used in the main experiment. This gave $r = 0{\cdot}075$, whilst from the experiment $r = 0{\cdot}095$.[a]

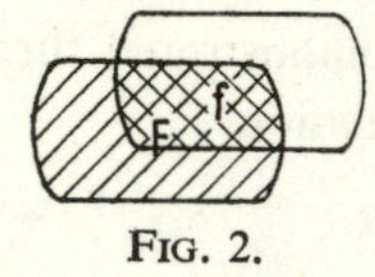

FIG. 2.

That the experimental value is greater than the theoretical is in accord with the fact that the intensity is greatest in the vertical direction, favoured by the counter disposition (section 7).[b] Clearly the coincidence rate can only attain the value given by this expression if the counter responds to each ray entering it with well-nigh 100% probability.[10] This is only possible for a corpuscular radiation, and not for γ-radiation. The radium experiment described above (section 8[c]) demonstrates directly the truth of this assertion for radioactive γ-rays. Since it may be questioned whether this conclusion may be extended without more ado to a radiation of the hardness of the high-altitude radiation, this point will be discussed more thoroughly.

A γ-ray can only produce a coincidence in the following way: the γ-ray, through two successive Compton processes, releases two secondary electrons, each of which enters one of the counters. To evaluate

[10] In this we agree with coincidence experiments which Geiger and Müller have made especially to investigate this question (*Forsch. & Fortschr.*, in the press).

the frequency of coincidences arising in this way, we consider the following schematic case (Figure 3): the γ-ray passes through the shield, from which an electron is released in the direction of the upper counter, and then the layer of absorber, where a second electron is produced in the direction of the lower counter. It is well known that the second-

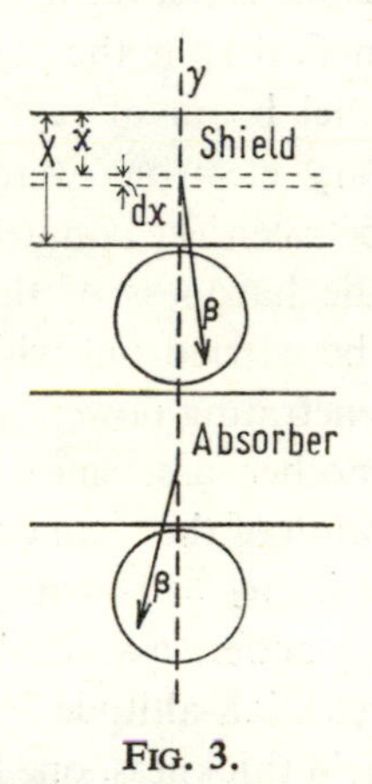

FIG. 3.

ary radiation of such short-wavelength γ-rays is predominantly directed forwards. Let μ be the scattering coefficient of the γ-rays and α the absorption coefficient of the secondary β-rays in the shield. The probability that a γ-ray, incident vertically on the counters, produces a secondary electron at a depth x in the shield, in an interval dx, is $\exp(-\mu x).\mu.dx$. The probability that this electron also emerges from the shield is $\exp(-\alpha(X-x))$ times as large, where X is the thickness of the shield. Thus the total probability that a γ-ray ejects an electron into the upper counter is

$$w_1 = \int_0^X \mu.e^{-\mu x-\alpha(X-x)}.dx = \frac{\mu}{\alpha-\mu}(e^{-\mu X}-e^{-\alpha X}).$$

A corresponding expression holds for the absorber, from which the probability of the appearance of the second electron is w_2. The probability for the occurrence of a coincidence is the product $w_1 w_2$, hence the coincidence fraction referred to the lower counter is $w_1w_2/w_2 = w_1$. w_1 would thus have to be nearly unity if one were to explain the observed coincidences through γ-rays in the way described. In the case of RaC γ-radiation $\alpha \gg \mu$; accordingly μ/α is obtained as an upper limit for w_1, which is at most of the order of magnitude 1/100. In fact in the Radium experiment (section 8)[c] no observable "Ra-coincidences" remained, despite the particularly favourable geometrical conditions (parallel beam of rays), when the coincidences produced by the secondary electrons were suppressed by suitably thick layers of absorber between the counters.

For a γ-radiation of the hardness of the high-altitude radiation this evaluation can only be carried out with due caution, since one knows very little of the penetrating power of extremely fast electrons. However, in our case another assessment is possible. For $\alpha = 0$, that is without any absorption of the secondary electrons in the shield, one has as an upper limit, $w_1 = 1-\exp(-\mu X)$, which is simply the probability that a γ-ray suffers a scattering process in the shield. Taking now for the filtered high-altitude radiation $\mu/\varrho = 10^{-3}$[f] and $X = 100$ g/cm² for the shield thickness, one has $w_1 < 0{\cdot}1$. In our case one must certainly be far removed from this upper limit, in which absorption and angular deviation of the secondary electrons were not taken into account.

Besides, in the experiments on the upper floor the shield was removed, so there would be no possibility of γ-rays generating coincidences in the manner just considered. The possibility that a γ-ray produces secondary electrons in the air directly above the apparatus or in the walls of the counter itself is remote, as one finds, for instance, the probability for a 100 cm layer of air is only about 4×10^{-4}, and for 1 mm zinc of the tube wall 2×10^{-3}. The possibility that one or both of the secondary electrons producing coincidences originate in the side shielding walls also needs no further discussion, as the conditions are clearly still more unfavourable for this than for the covering shield. Besides, a more detailed inspection of the results (section 10) shows that the shield exerted no influence on the coincidences other than that corresponding to its absorbing power for the high-altitude radiation.

One can perhaps summarize the whole discussion in a single argument: the mean free path of a γ-ray between two electron-ejecting processes would be $1/\mu = 1000$ cm water $= 90$ cm lead $= 52$ cm gold for the filtered high-altitude radiation. Hence one sees at once that a quite exceptional accident must be supposed, if two electrons produced by the same γ-ray should display the necessary penetrating power and correct direction to strike both counters directly.

For completeness, there is still a theoretical possibility to be considered that for γ-rays of very short wavelength entirely new phenomena could appear, giving the γ-ray the properties of a corpuscular ray, so that our results might also be explained by γ-rays. For this it would be necessary that the path length between two scattering processes drop to a small fraction, not only of the quantum-mechanical, but also of the classical value ($1/\mu = 5$ g/cm^2), and that only an exceptionally small amount of energy be transmitted to the secondary electron at each scattering process. This *ad hoc* assumption cannot be justified at present from either the experimental or theoretical standpoints.

We conclude then, that *each coincidence signifies the passage of one and the same corpuscular ray through both counters*, and that we have measured the absorbability of these corpuscular rays by the decrease of the coincidences.

10. THE ABSORBABILITY OF THE CORPUSCULAR RAYS

This must now be compared with that of the high-altitude radiation itself. Figure 4 will serve for this, where the results of our observations from Tables 2 and 3 are summarized, in so far as they relate to the frequency of coincidences (o) and the discharges in the lower counter (x); the last are corrected, according to section 6, for residual and background radiation: the abscissa is the thickness of absorber, expressed in grams of lead/cm^2 corresponding to the electron content of the actual absorbers (Au, Pb, Fe, concrete). The abscissae 0 and 80 apply to the observations on the upper floor, without and with the gold block, with the abscissae 350 and 430 correspondingly for the ground floor. The observations on the upper floor hold the chief interest.

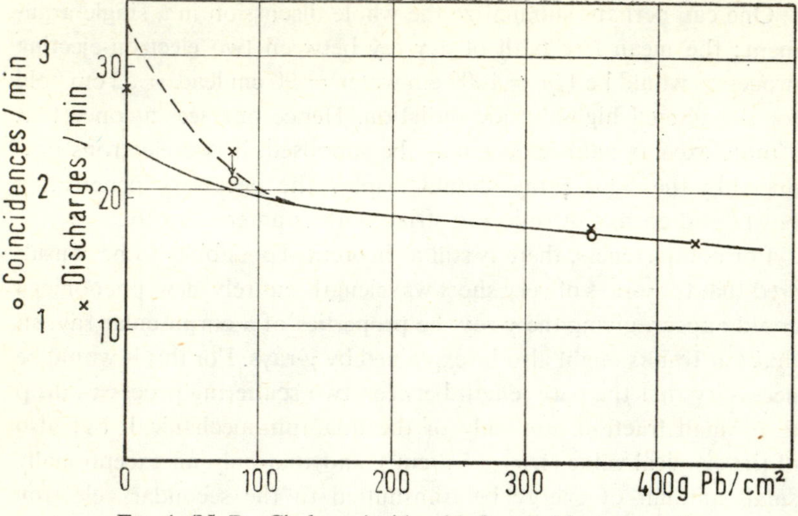

FIG. 4. (N. B.: Circles coincide with first and last crosses.)

One sees that the insertion of the gold block also depressed the discharges, but, as a percentage, about 1/3 less than the coincidences. However, it should be noted that the gold block covered the lower counter only very incompletely; that part of the high-altitude radiation entering obliquely through the side shield was not caught by the absorber. From the geometrical conditions (Figure 1) and from consideration of the anisotropy of the high-altitude radiation, one would assess this part as making up about 1/3 of the whole radiation acting on the counter. If one thus relates the decrease in the discharges (just as that of the coincidences) solely to the radiation actually penetrating the gold block, one must lower the second *x*-point corresponding to the discharges, and it then falls at least close to the corresponding *o*-point of the coincidences: *viz. the corpuscular rays are just as strongly absorbed as the high-altitude radiation itself under the same conditions.* The observations on the ground floor cannot themselves be discussed in the same way, as the differences with and without the gold block are too small and so too uncertain, but they are still in accord with our conclusion.

It may at first seem remarkable that filtering through only 30 cm of Pb-equivalent suffices to decrease the originally considerable absorbability of the high-altitude radiation to a small fraction. However, this is in complete agreement with what is found by other methods, and may arise from the special condition which, according to the observations of Hoffmann and collaborators as well as Myssowsky and Tuwim, are found on the transition of high-altitude radiation between two different absorbers. That is, if one lets the radiation from air enter a material of higher atomic mass, its absorbability seems at first anomalously high, and then after a certain depth of penetration falls back to its normal value. In our experiments the conditions for the appearance of this transition effect were present in the upper floor, but not on the ground floor, where the heavy-atom shield was inserted befor- the gold absorber.

* * * * *

In Figure 4 the absorption curve for high-altitude radiation taken from the work of Steinke is entered, specifically that in lead (full line and that for the passage from lead from air (dashed). The agreemene is very satisfactory, when one remembers that our geometrical condit tions were not quite identical with those of Hoffmann and Steink).

* * * * *[d]

V. Properties of the High-altitude Radiation

The probable characteristics of such a high-energy corpuscular radiation can only be briefly sketched, for it being certainly a question of the distant extrapolation of the familiar, such considerations will as yet have a very hypothetical character.

12. INTENSITY

If one counts 28 deflections per minute = 0·5/sec in the lower counter for the unfiltered high-altitude radiation, considered as isotropic, and if $N.d\sigma.d\Omega$ is the number of particles per second in a beam of cross-section $d\sigma$ and opening angle $d\Omega$, one has that $0{\cdot}5 = N\int F.d\Omega$, where F denotes the projected area of the counter in the direction ϑ, φ, $d\Omega$ denotes the elementary cone, and the integral extends over a hemisphere; the angle ϑ is measured from the vertical (see section 9). This yields for the number of particles per second on a square centimetre of the Earth's surface

$$N\int \cos\vartheta.d\Omega = 0{\cdot}5\,\frac{\int \cos\vartheta.d\Omega}{\int F.d\Omega} = 5\times 10^{-3}.$$

The actual anisotropy of the radiation will have the effect of increasing the theoretical expression, so that one can reckon with about 1/100 particles/cm^2 sec.[11]

This value is so remarkably small that it seems almost hopeless to observe the charge of an isolated body under the influx of these rays.

* * * * *

These corpuscular rays can thus be of no importance in the maintenance of the Earth's negative charge.

13. IONIZING POWER

If one takes the ionizing strength of the unfiltered high-altitude radiation at sea level as 1·5 J (ions/cm^3 sec at normal pressure), for the volume of a counter it gives 300 J. The number of particles detected by the counter amounted to 0·5/sec, so on average a particle must produce 600 ion pairs. One can take about 7 cm as the mean path length of a particle in the counter. This gives $i = 90$ ion pairs/cm of normal air for the differential ionizing power of the corpuscular rays. For fast β-particles it has been found that i seems to approach a limiting value of about 40.[12] It is certainly very striking that this number

[11] D. Skobelzyn *(loc. cit.)* finds about 0·02/cm^2 sec with the Wilson method.
[12] Cf. *Handb. d. Phys.* **XXIV**, 53 (1927).

coincides in order of magnitude with that estimated here for the high-altitude radiation; Skobelzyn also draws the same conclusion from his Wilson photographs of the high-altitude radiation particles. Nevertheless, our value appears to be distinctly greater than that for fast β-particles.[e]

14. ENERGY, ABSORPTION AND SCATTERING

For corpuscular rays of such enormous penetrating power it is probably hardly possible to deduce the energy directly from the formal absorption coefficient. This is indeed ruled out in the energy range already known for nuclear rays (proton- or α-rays), which travel on almost perfectly rectilinear paths through matter: here one must reckon with range instead of absorption coefficient. Practically the same is true of very energetic electrons. ... Bohr's slowing down formula for electrons reads, with the numerical factor of Varder and Schonland, for high energies ... (V = energy in eVolt), roughly

$$R\varrho = 10^{-6}\,\mathrm{V}.\ (R\text{: range; } \varrho\text{: density}) \tag{4}$$

* * * * *

One can now reach formula (4) in an entirely different way. If one accepts that the limiting value hitherto reached for the differential ionization by electrons ($i = 40$) holds right up to the limit $v = c$, and that the energy loss on an ion pair also remains at its known value of around 30 eV, the energy loss for 1 cm of normal air comes to $40\times30 = 1200$ eV, or 10^6 eV for a layer of 1 g/cm^2. If one neglects the final short part of the path where the velocity falls well below the velocity of light, and gross deflection occurs, the result is identical with equation (4). These considerations seem to give each other some support.[e]

Now, one can accept with some certainty that the high-altitude radiation has its origin outside the confines of the Earth's atmosphere (see further below). To reach sea level it has to penetrate 10^3 g/cm^2 at vertical incidence. Hence at the top of the atmosphere $R\varrho$ is in any case $> 10^3$, and consequently $V > 10^9$ eV In fact much higher values have to be reckoned with. For as Regener has recently been able to follow the high-altitude radiation as deep as 230 m in water, this already gives $R\varrho = 2{\cdot}4\times10^4$, $V = 2{\cdot}4\times10^{10}$ eV, for vertical passage, provided that it still behaves as a corpuscular radiation at these depths.[13] For oblique paths of the rays, correspondingly higher values obtain. The values become still greater if the ionizing power exceeds $i = 40$, as appears to be the case in our experiments and is expected for positive rays (proton or α-rays).

15. NATURE AND ORIGIN OF THE RADIATION PARTICLES

So far, one has only been able to make conjectures about the nature of the particles which form the high-altitude radiation, as it is not easy to decide this experimentally. It is to be borne in mind that at such high energies as are involved here electrons and protons approach each other more and more in their behaviour (apart from sign of charge).

* * * * *

As for the deflection in a magnetic field, the question arises of experiments outside the laboratory concerning the effect of the Earth's magnetic field. From the theory of the aurora borealis of Birkeland, Störmer, Lenard *et al.*[14] it is known that by virtue of its huge extent the Earth's magnetic field, despite its small field strength, exerts an important effect even on cathode rays of the energy in question here. Thus, if such a radiation came from the external universe its intensity at the Earth's surface would depend on the magnetic latitude; amongst other things there would be a latitude zone in which a particularly high intensity appeared. From this point of view the relation of the high-altitude radiation to the isochasms[15] (lines of equal frequency of the

[13] E. Regener, *Die Naturwissensch.* 17, 183 (1929); is there possibly in this hard residual radiation a γ-bremsstrahlung produced by the high-altitude radiation?

[14] Cf. Müller-Pouillet, *Lehrb. d. Phys.* V, 493 (1928) (Angenheister article).

[15] W. Kolhörster, *Die Naturwissensch.* 7, 412 (1919); cf. also R. Swinne, *ibid.* 7, 529 (1919).

northern lights) gains heightened interest. Measurements concerning this are intended. If this gives a distribution of the kind described it will yield a further argument for the corpuscular nature of the high-altitude radiation and moreover give a clearer indication of its extra-terrestrial origin. Recent measurements by Clay[16] perhaps give indications of such an effect.

* * * * *

The absence of a cycle in solar time and the lack of an eclipse effect still argue against a solar origin, though because of the Earth's magnetic deviation such dependences need not be so sharp as for a γ-radiation.

Although the question of the origin of the high-altitude radiation still finds no answer through our results, the whole matter must now be discussed afresh. As long as one adheres to a γ-character for the high-altitude radiation, one must almost inevitably think of a production process on the atomic scale. A corpuscular radiation, on the other hand, could acquire its energy in very weak but exceedingly extensive fields of force; since for example the distance of the "extra-galactic nebulae" is reckoned in millions of light years according to modern ideas.

We have particularly to thank the Notgemeinschaft der Deutschen Wissenschaft for providing a grant.

Charlottenburg, May 1929.

Notes

[a] We omit several sections of this long paper, including investigation of the residual radiation in the counters, and including

[b] the effect of altering the counter separation—showing that more rays come from near the vertical than from other directions;

[c] generation of coincidences by γ-rays from RaC—very inefficient and cut considerably by 1 mm aluminium;

[d] a theoretical calculation of the ranges of secondary β-rays.

[e] The total ionization for $v \doteqdot c$ is in fact nearer to B. and K.'s value.

[f] The mean free path for Compton scattering is not much less than that for absorption of γ-rays, as the energy transfer is large.

[16] J. Clay, *Amst. Proc.* **31**, 1091 (1928).

3. The Earth-magnetic Effect and the Corpuscular Nature of (Cosmic) Ultra-radiation. IV†

J. Clay

§ 1. In our previous communications[1] (II and III) the variation of intensity of the ultra-rays on the Earth's surface was given. The true value of the effect was yet subject to some doubt, since others did not find a variation in other parts of the Earth. Millikan[2] found nearly no variation between Pasadena and Churchill. Bothe and Kolhörster[3] only found a doubtful variation between Hamburg and Spitzbergen. Corlin[4] has given a discussion on this subject.

The observations to be discussed in the present paper were performed with two instruments constructed by Dr. Steinke. They were first compared at Amsterdam. Then one of the two was mounted on board the motorship *Christiaan Huygens*, where the observations were made by Dr. Berlage during the journey from Genoa to Batavia. The instrument consisted of an ionization chamber of 22 liters filled with carbon dioxide under a pressure of 11 atmospheres and surrounded by a shield of 13 cm iron. The wall of the chamber is at a potential of 130 volts. The needle in the ionization chamber is connected with an electrometer. The ionization current, which thus charges up the electro-

† *Koninklijke Akademie van Wetenschappen te Amsterdam, Proceedings of the Section of Sciences*, **35**, 1282–1290 (1932). (Communicated by Prof. P. Zeeman. Communicated at the meeting of December 17th, 1932.)

[1] *Proc. Roy. Acad. Amsterdam*, **XXXI**, 1091 (1928); **XXXIII**, 711 (1930). *Die Naturw*. **20**, 687 (1932).

[2] R. Millikan, *Phys. Rev*. **36**, 1596 (1930).

[3] W. Bothe and W. Kolhörster, *Berl. Ber*. **26** (1930).

[4] A. Corlin, *Lund. Obs. Circ*. **1**, 3 (1931).

meter, is gradually compensated by a known charge of opposite sign induced on it. The sensitivity of the electrometer was such that, if the charge had not been compensated, the deflection would have been a hundred and eighty scale divisions for 1 hour. At the end of every hour, the needle is photographed, so that its position may be read from the photograph with an accuracy of one-tenth of a division. The measurements concerning the instrument may be considered accurate to 1°/₀₀. With this instrument continuous 1 hour observations were performed, the result of which is given in the graph (Figure 1).

The falling off of intensity confirms our observations of the previous voyages. However the present observations are much more accurate and more detailed. Point *A* is the mean of 100 observations at Amsterdam, *B* the mean of our previous observation at Bandoeng reduced to sea level. The values are plotted according to magnetic coordinates,[5] and measured from the reduced magnetic North pole (78° N.L. and 69° W.L.). That the minimum of our curve deviates a little from the magnetic equator thus defined may be due to the fact that the dipole in the Earth does not quite coincide with the Earth's centre. In reality the acline is more to the South.

Points from *G* to *P* are 6 hour means, from *P* to *S* 12 hour means. In the latter case there were over 200 hour observations, therefore the values are very accurate (there is only a barometric influence between Suez and Batavia of 0·2%).

Combining these results with those of Millikan and of Bothe and Kolhörster and with the recent data of Bennett[6] and his collaborators —the latter reduced to our initial value—because their shield had a different thickness, we obtain an intensity curve, which runs horizontal from the Pole to *A*, then diminishes towards the magnetic equator by 16% and after crossing this equator again increases towards the South Pole, according to Compton's observations[7] (Figure 2).

We may conclude from this that the ultra-radiation is a charged corpuscular radiaton as we shall show more in detail in the following.

[5] J. Bartels, *Handbuch der Experimental Physik*, **25**[1], 586.
[6] R. D. Bennett, *C. S. Phys. Rev.* **42,** 446 (1932).
[7] A. H. Compton, *Phys. Rev.* **41,** 681 (1932).

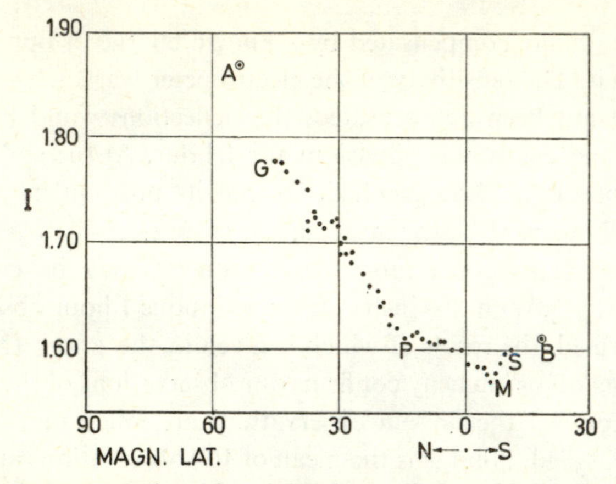

FIG. 1. Variation of the intensity of the ultra-radiation with the Earth magnetic latitude.

§ 2. Let us first consider the balance of arguments for and against this conception:

1. The absorption experiments by Bothe and Kolhörster[8] give arguments in favour of corpuscular rays, but do not exclude the possibility that the rays they have measured would be secondary corpuscular radiation from primary gamma rays. Rossi's experiments would moreover be against monochromatic gamma radiation.[9]

Of the fact that the absorption coefficient for these counter experiments has the same value as for the primary radiation Johnson[10] has given a beautiful explanation.

2. Regener's[11] absorption measurements in great depths of water suggest gamma radiation because of the character of the intensity decrease, but the possibility of corpuscular radiation is not excluded, since corpuscular rays with a Maxwellian distribution of energy might show a similar decrease of intensity at great depths.

3. Regener's[12] measurements at high altitude, where it is found

[8] W. Bothe and W. Kolhörster, *Z. f. Phys.* **56**, 751 (1929).
[9] Rossi, *Phys. Rev.* **36**, 606 (1930).
[10] Thomas H. Johnson. *Phys. Rev.* **41**, 545 (1932).
[11] E. Regener, *Die Naturw.* **17**, 183 (1929). *Z. f. Phys.* **74**, 433 (1932).
[12] E. Regener, *Die Naturw.* **20**, 695 (1932).

that at 2·5 cm of mercury the ionization still increases are against the conception of monochromatic primary gamma radiation producing corpuscular rays of high range, coming from the outside of the atmosphere.

4. Street and Johnson's[13] experiments with double and triple coincidences in connection with absorption in lead strongly point to corpuscular rays and are against gamma rays.

5. The influence of the earth-magnetic field decidedly shows the corpuscular nature of the rays and is against primary gamma radiation, since for a noticeable effect the paths of the corpuscular rays must be so long, that it seems impossible that they have originated from primary gamma rays at such a high altitude above the Earth.

Until recently there was some doubt as to whether the corpuscular rays could have an energy sufficient to pass through the Earth's atmosphere. But the experiments by Skobelzyn,[14] by Millikan and Anderson[15] and by Kunze[16] show that even the secondary rays (e.g. for pairs originating from the wall of the expansion chamber) may have energies of 10^9 e. volt.

One may briefly summarize the situation as follows:

	gamma	corpuscular
1. Absorption measurements Bothe, Kolhörster, Rossi		for
2. Absorption measurements Regener in water	for	
3. Ionization curve in upper atmosphere to 28 km Regener	against	
4. Measurements by Street and Johnson with double and triple coincidences	against	for
5. Influence of earth-magnetic field on intensity	against	for

§ 3. Assuming now *a system of corpuscular rays of all energies from zero to infinity to be incident on the Earth*, then there are two limitations

[13] J. C. Street and Th. J. Johnson, *Phys. Rev.* **42,** 142 (1932).

[14] Skobelzyn, *Z. f. Phys.* **54,** 686 (1929).

[15] R. Millikan and C. Anderson, *Phys. Rev.* **39,** 325 (1932). C. Anderson, *Phys. Rev.* **41,** 321 (1932).

[16] Paul Kunze, *Z. f. Phys.* **79,** 203 (1932).

for incidence on the Earth's surface at the bottom of the atmosphere: in the first place the limitation imposed by Störmer[17] forbidden spaces, and in the second place the limitation imposed by the thickness of the atmosphere, necessitating for the primary radiation a lower limit corresponding to the loss of energy through the atmosphere. Below this limit no radiation will reach the Earth's surface.

In order to determine, how large the energy of the primary corpuscles must be, if they are to reach the Earth's surface through the atmosphere, we may follow different methods.

1. From Johnson's[18] calculations based on Schindler's[19] determinations of transition effects follows that in air there are on the average 2·5 secondary corpuscular rays (secondaries) in equilibrium with every primary. The energy of these secondaries is 1·3 times the energy of secondaries in iron. In iron the energy of secondaries is, on the *average*, $2{\cdot}5 \times 10^8$ e. volt, according to Anderson's photographs, where one has to consider especially the pairs which certainly originate in the iron walls of the chamber. Hence the secondaries in air would have an energy of $3{\cdot}2 \times 10^8$ e. volt. If this energy be used for ionization, this requires 1200 e. volt per cm.

These rays may therefore pass through $2{\cdot}6 \times 10^5$ cm of air. For the entire atmosphere of the hight $8{\cdot}0 \times 10^5$ cm, reduced to 76 cm of mercury, this amounts to $2{\cdot}5 \times \frac{8{\cdot}0}{2{\cdot}6} \times 3{\cdot}2 \times 10^8 = 2{\cdot}5 \times 10^9$ e. volt for the total energy of the secondaries accompanying one primary. To this is to be added the direct ionisation energy of the primary through the atmosphere, which is $1{\cdot}0 \times 10^9$ e. volt. The total *minimum* energy lost directly and indirectly by one primary passing through the atmosphere is therefore $3{\cdot}5 \times 10^9$. In order to account for the fact that primaries are not only incident in the vertical but also in other directions one must take a factor of about 1·6, and thus the necessary energy would be on *the average* $5{\cdot}6 \times 10^9$ e. volt.

[17] Carl Störmer, *Ergebnisse der Kosmischen Physik*, **I**, 1–86 (1932). W. Heisenberg, *Ann. d. Phys. 5ᵉ F.* **13**, 430 (1932).

[18] Th. H. Johnson, *Phys. Rev.* **41**, 545 (1932).

[19] H. Schindler, *Z. f. Phys.* **72**, 625 (1931).

2. One may also carry out an estimate according to a different method, making use of the impulses of ionization occurring in ionization chambers according to observations by Steinke and Schindler[20] and by Messerschmidt[21]. They find that from lead corpuscles are emitted with an energy of 10^9 e. volt and perhaps still more. They further state, that after covering up with 10 cm of lead, an additional 10 cm of lead still gives an increase in the number of impulses, hence one may conclude that the range of the corpuscles under consideration is between 10 and 20 cm of lead.

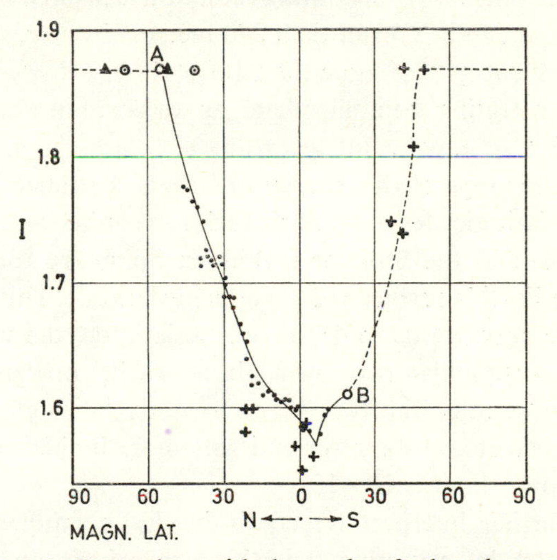

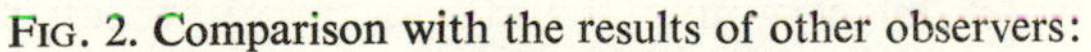

FIG. 2. Comparison with the results of other observers:
+ Observations of A. H. Compton.
▲ Observations of W. Bothe and W. Kolhörster.
⊙ Observations of R. Millikan and Cameron.

This estimate is further confirmed by our previous measurements[22] on absorption in lead, which show that the *local radiation*, which we consider as secondary radiation from the air (because radioactive

[20] E. Steinke and H. Schindler, *Z. f. Phys.* **75,** 115 (1932).
[21] W. Messerschmidt, *Z. f. Phys.* **78,** 78 (1932).
[22] *Proc. Roy. Acad. Amsterdam,* **XXXI,** 1091 (1928).

radiations were cut off by these experiments), is screened off by about 15 cm of lead.

Assuming that the impulses of 10^9 e. volt originate in the inner surface layer of the lead, the corresponding sets of corpuscles will have a range of about 20 cm of lead or 232 cm of water, so that 10 meters of water would correspond to $4{\cdot}2\times10^9$ e. volt. For a corpuscle to penetrate the atmosphere would therefore require a minimum energy of $4{\cdot}2\times10^9$ e. volt, and, taking different directions, an average energy of $6{\cdot}7\times10^9$ e. volt.

We know that the most penetrating radiation measured by Regener[23] is still present at depths greater than 240 meters of water, so that these rays must have energies of more than 100×10^9 e. volt.

The part penetrating the atmosphere, as we saw, must have energies of at least $3{\cdot}5\times10^9$ e. volt. But according to Störmer's theory the forbidden space for rays of this energy intersects the Earth's surface at 46° magnetic latitude. Rays of all directions require, on the average, 6×10^9 e. volt, and the Störmer forbidden boundary for these rays intersects the Earth's surface at 34° magnetic latitude. This means that from the pole downwards to 46° magnetic latitude, the whole of the spectrum of corpuscular rays from about 4×10^9 onward to higher energies will penetrate. But from 46° on towards the magnetic equator, the energy spectrum, at its lower end gets more limited, until, at the equator the lower limit is 10×10^9.

We may further interpret Regener's hardest component under 80 meters of water by assuming that these rays have an exponential distribution of energy, since the absorption curve gives a true picture of the spectrum of ranges, hence of the energy spectrum, if energy is assumed to be proportional to range. An exponential absorption then corresponds to an exponential distribution of energy.

Thus an ionization curve $e^{-\mu x}$, where the absorption coefficient, according to Regener[24] has the value $= 0{\cdot}020$ per meter of water, leads to an energy distribution $e^{-\frac{\varepsilon}{3\times10^{10}}}$ as follows from the result just

[23] E. Regener, *Z. f. Phys.* **74**, 433 (1932).
[24] E. Regener, *ibid.*

obtained that a range of 10 meters of water corresponds to an energy of 6×10^9 e. volt.

§ 4. We now obtain the following picture. We plot the spectrum of these rays in $N = N_0 e^{-\frac{\varepsilon}{3\times10^{10}}}$ in Figure 3, where N is plotted vertically and ε horizontally. The part of the spectrum *ABC*, cut off at *A*, occurs about uniformly from the magnetic pole to 46°, so that between the pole and 46° a very little change of intensity of ionization is to be ex-

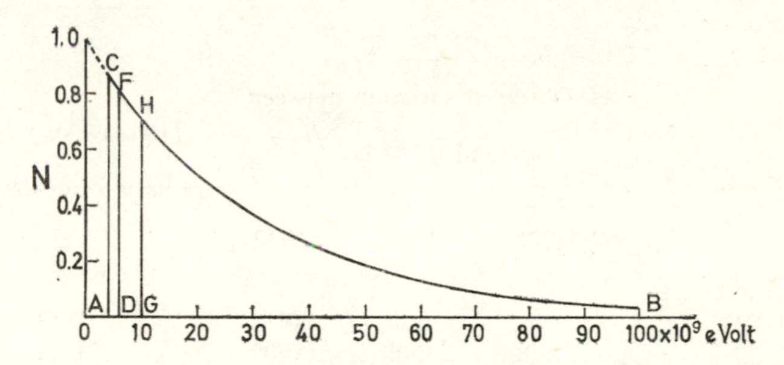

FIG. 3. Distribution of the number of corpuscles over their energy and the variation of their ionization with magnetic latitude. The part of the graph between *O* and *A* will be discussed later.

pected, as is confirmed by many measurements, as indicated above. At 46° the Störmer boundary for 4×10^9 e. volt begins. Between 46° and 34° the rays between 4×10^9 and 6×10^9 e. volt, which may reach the Earth's surface from various directions of space, will gradually be eliminated on account of Störmer's limitation. Proceeding from 34° further towards the magnetic equator, rays of energy larger than 6×10^9 are gradually eliminated, until at the equator everything below 10×10^9 e. volt has been cut off.

Calculating the number of particles which is eliminated between 34° northern magnetic latitude and equator, one obtains a decrease of 12% (area *DFHG*), whereas our observations yield a decrease of 9% (one may expect the ionization per cm³ to be proportional to the number of corpuscular rays left). For the decrease from 46° to the equator (energy limit from 4×10^9 to 10×10^9) we calculate a decrease of 18% (area *ACHG*), whereas our observations give a decrease in ionization of

12%. Our observations of 1928 have given 12% and in 1929 11%. A better agreement could hardly be expected, the more since in the simple picture followed here, the distribution outside Störmer's forbidden space for a certain energy has been assumed to be homogeneous, which does not follow from Störmer's theory, but is intended only as a rather rough approximation.

We may summarize the results in Table 1.

TABLE 1

Magn. lat. φ	Observed variation between $\varphi°$ and 0° M.L. $\frac{dI}{I_\varphi}$			Decrease of *primary* radiation calculated $\frac{dI}{I_\varphi}$
	1928	1929	1932	
43°	0·12	0·11	0·12	0·18
33°	0·09	0·08	0·09	0·12

It may be expected that the observed difference would have been somewhat larger, if we could have observed at the points of maximum horizontal intensity, which occurs in the South Chinese Sea near Indo-China. We intend to pursue the same calculation with Maxwellian distribution which will be especially for the lowest energies of another and more probable form.

The mean energy to be expected at the Earth's surface (lower limit 6×10^9 e. volt) according to the spectral distribution of Figure 3 is calculated 30×10^9 e. volt, whereas Johnson concludes from Anderson's experiments to a mean energy of 22×10^9 e. volts for a primary ray, which therefore outside the atmosphere would have been 28×10^9 e. volt.

§ 5. In conclusion, it will be clear that, if the above explanation is correct, the radiation will gradually become harder, as one passes from 46° latitude towards the magnetic equator. Indeed there are in my former observations three cases which show this. In the voyages of

1928 and 1929, the 8 cm iron shield surrounding electrometer was removed now and then and the difference was measured. This difference always increased with the distance from the equator.[25]

	dI (1928)	*dI′* (1929)	magn. latitude
Suez canal	0·48	0·36	29°
Red Sea, North		0·31	21°
Red Sea, South	0·32		15°
Gulf of Aden	0·18		8°
Indian Ocean		0·26	6°

The values, though not accurate, certainly show the variation expected.

The third case is the following. In 1927 Büttner[26] has measured the absorption for 3 cm of lead with a Kolhörster apparatus at an altitude of 3470 meters. It is a curious coincidence that I happened to measure in the same year the absorption also for 3 cm of lead at 3024 meters with an identical apparatus. Büttner found $17{\cdot}4\times10^{-3}$ cm^{-1} of lead at about 50°N magnetic latitude, and I obtained $11{\cdot}8\times10^{-3}$ cm^{-1} at 18°S magnetic latitude.

It is to be regretted that the absorption measurements in water just published by Benade[27] cannot be compared with those of Regener, since they were made at different altitudes. At any rate it may be seen that the upper parts of their curves do not coincide, but the lower parts do, as is to be expected.

§ 6. We have reached the conclusion *that the ultra-radiation incident on the Earth is a charged corpuscular radiation, in all probability of a Maxwell distribution of which the hardest end is approximately exponential with a mean energy of about 3×10^{10} e. volt, which is cut off at the lower limit of 4×10^{9} e. volt by the atmosphere, whereas between 50° magnetic latitude and the magnetic equator an additional 16% is cut off at the lowest side in consequence of Störmer's forbidden spaces.*

[25] *Proc. Roy. Acad. Amsterdam*, **XXXI,** 1091 (1928); **XXXIII,** 711 (1930).
[26] K. Büttner, *Z. Geophys.* **3,** 161–237 (1927).
[27] J. Benade, *Phys. Rev.* **42,** 290 (1932).

In the atmosphere this primary radiation produces a secondary radiation of positive rays (protons and perhaps neutrons) of great energy and of negative rays and some of these rays will give in their turn a tertiary radiation. The paths of these rays depend on the pressure of the air.

On this base it is possible also to give an explanation of the ionization curve of Regener[28] in the stratosphere and of the high conductivity of the upper layers of the atmosphere, which I will give in a following paper.

The interpretation of a Maxwell distribution being correct, one might ask, what is the source of such an enormous electron temperature, a problem which might be of interest for astrophysics. I suppose they must originate from the hot stars of the Milky system.

It will be of much value to investigate more in detail the intensity and range distribution from about 50° latitude to the equator and, especially, to measure the ionization in the upper atmosphere according to Regener's method.

A determination of the direction from which the primary rays originate will involve serious difficulties on account of the appreciable curvature of the rays in the earth-magnetic field. Only the rays with highest energies will be suited for this purpose.

In conclusion we wish to express our cordial thanks to Dr. Berlage for his careful help for the measurements and to Dr. H. Zanstra for valuable help in various fruitful discussions and calculations. Further to the Direction and Inspection of the Steamship Company "Nederland" for their readiness in giving facilities and to Captain Mörzer Bruins, First Officer Steen and members of the crew of the M.S. *Christiaan Huygens* for their active help.

Natuurkundig Laboratorium, Amsterdam, December 10, 1932

[28] E. Regener, *Die Naturw.* **20,** 695 (1932).

4. The Positive Electron[†]

CARL D. ANDERSON[‡]

ON August 2, 1932, during the course of photographing cosmic-ray tracks produced in a vertical Wilson chamber (magnetic field of 16,000 gauss) designed in the summer of 1930 by Professor R. A. Millikan and the writer, the tracks shown in Figure 1 were obtained, which seemed to be interpretable only on the basis of the existence in this case of a particle carrying a positive charge but having a mass of the same order of magnitude as that normally possessed by a free negative electron. Later study of the photograph by a whole group of men of the Norman Bridge Laboratory only tended to strengthen this view. The reason that this interpretation seemed so inevitable is that the track appearing on the upper half of the figure cannot possibly have a mass as large as that of a proton for as soon as the mass is fixed the energy is at once fixed by the curvature. The energy of a proton of that curvature comes out 300,000 volts, but a proton of that energy according to well established and universally accepted determinations[1] has a total range of about 5 mm in air while that portion of the range actually visible in this case exceeds 5 cm without a noticeable change in curvature. The only escape from this conclusion would be to assume that at exactly the same instant (and the sharpness of the tracks determines that instant to within about a fiftieth of a second) two independent

[†] *Phys. Rev.* **43**, 491–494 (1933). (Received February 28th, 1933.)

[‡] California Institute of Technology, Pasadena, California.

[1] Rutherford, Chadwick and Ellis, *Radiations from Radioactive Substances*, p. 294. Assuming $R \propto v^3$ and using data there given the range of a 300,000 volt proton in air S.T.P. is about 5 mm.

electrons happened to produce two tracks so placed as to give the impression of a single particle shooting through the lead plate. This assumption was dismissed on a probability basis, since a sharp track of this order of curvature under the experimental conditions prevailing occurred in the chamber only once in some 500 exposures, and since there was practically no chance at all that two such tracks should line up in this way. We also discarded as completely untenable the assumption of an electron of 20 million volts entering the lead on one side and coming out with an energy of 60 million volts on the other side. A fourth possibility is that a photon, entering the lead from above, knocked out of the nucleus of a lead atom two particles, one of which shot upward and the other downward. But in this case the upward moving one would be a positive of small mass so that either of the two possibilities leads to the existence of the positive electron.

In the course of the next few weeks other photographs were obtained which could be interpreted logically only on the positive-electron basis, and a brief report was then published[2] with due reserve in interpretation in view of the importance and striking nature of the announcement.

Magnitude of Charge and Mass

It is possible with the present experimental data only to assign rather wide limits to the magnitude of the charge and mass of the particle. The specific ionization was not in these cases measured, but it appears very probable, from a knowledge of the experimental conditions and by comparison with many other photographs of high- and low-speed electrons taken under the same conditions, that the charge cannot differ in magnitude from that of an electron by an amount as great as a factor of two. Furthermore, if the photograph is taken to represent a positive particle penetrating the 6 mm lead plate, then the energy lost, calculated for unit charge, is approximately 38 million electron-volts, this value being practically independent of the proper mass of the

[2] C. D. Anderson, *Science*, **76**, 238 (1932).

particle as long as it is not too many times larger than that of a free negative electron. This value of 63 million volts per cm energy-loss for the positive particle it was considered legitimate to compare with the measured mean of approximately 35 million volts[3] for negative electrons of 200–300 million volts energy since the rate of energy-loss for particles of small mass is expected to change only very slowly over an energy range extending from several million to several hundred million volts. Allowance being made for experimental uncertainties, an upper limit to the rate of loss of energy for the positive particle can then be set at less than four times that for an electron, thus fixing, by the usual relation between rate of ionization and charge, an upper limit to the charge less than twice that of the negative electron. It is concluded, therefore, that the magnitude of the charge of the positive electron which we shall henceforth contract to positron is very probably equal to that of a free negative electron which from symmetry considerations would naturally then be called a negatron.

It is pointed out that the effective depth of the chamber in the line of sight which is the same as the direction of the magnetic lines of force was 1 cm and its effective diameter at right angles to that line 14 cm, thus insuring that the particle crossed the chamber practically normal to the lines of force. The change in direction due to scattering in the lead,[3] in this case about 8° measured in the plane of the chamber, is a probable value for a particle of this energy though less than the most probable value.

The magnitude of the proper mass cannot as yet be given further than to fix an upper limit to it about twenty times that of the electron mass. If Figure 1 represents a particle of unit charge passing through the lead plate then the curvatures, on the basis of the information at hand on ionization, give too low a value for the energy-loss unless the mass is taken less than twenty times that of the negative electron mass. Further determinations of $H\varrho$ for relatively low energy particles before and after they cross a known amount of matter, together with a study of ballistic effects such as close encounters with electrons, involv-

[3] C. D. Anderson, *Phys. Rev.* **43**, 381A (1933).

ing large energy transfers, will enable closer limits to be assigned to the mass.

To date, out of a group of 1300 photographs of cosmic-ray tracks 15 of these show positive particles penetrating the lead, none of which can be ascribed to particles with a mass as large as that of a proton, thus establishing the existence of positive particles of unit charge and of mass small compared to that of a proton. In many other cases due either to the short section of track available for measurement or to the high energy of the particle it is not possible to differentiate with certainty between protons and positrons. A comparison of the six or seven hundred positive-ray tracks which we have taken is, however, still consistent with the view that the positive particle which is knocked out of the nucleus by the incoming primary cosmic ray is in many cases a proton.

From the fact that positrons occur in groups associated with other tracks it is concluded that they must be secondary particles ejected from an atomic nucleus. If we retain the view that a nucleus consists of protons and neutrons (and α-particles) and that a neutron represents a close combination of a proton and electron, then from the electromagnetic theory as to the origin of mass the simplest assumption would seem to be that an encounter between the incoming primary ray and a proton may take place in such a way as to expand the diameter of the proton to the same value as that possessed by the negatron. This process would release an energy of a billion electron-volts appearing as a secondary photon. As a second possibility the primary ray may disintegrate a neutron (or more than one) in the nucleus by the ejection either of a negatron or a positron with the result that a positive or a negative proton, as the case may be, remains in the nucleus in place of the neutron, the event occurring in this instance without the emission of a photon. This alternative, however, postulates the existence in the nucleus of a proton of negative charge, no evidence for which exists. The greater symmetry, however, between the positive and negative charges revealed by the discovery of the positron should prove a stimulus to search for evidence of the existence of negative protons. If the neutron should prove to be a fundamental particle of a new kind

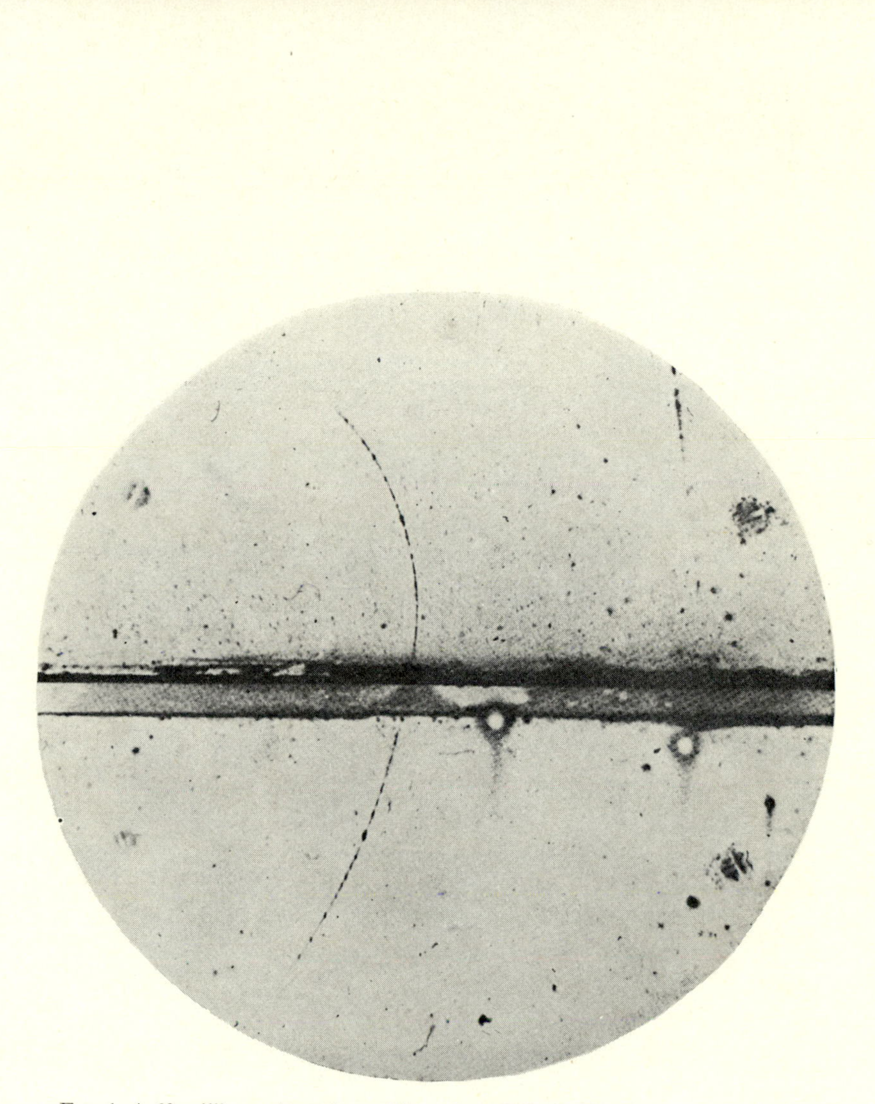

FIG. 1. A 63 million volt positron ($H\varrho = 2{\cdot}1\times10^5$ gauss-cm) passing through a 6 mm lead plate and emerging as a 23 million volt positron ($H\varrho = 7{\cdot}5\times10^4$ gauss-cm). The length of this latter path is at least ten times greater than the possible length of a proton path of this curvature.

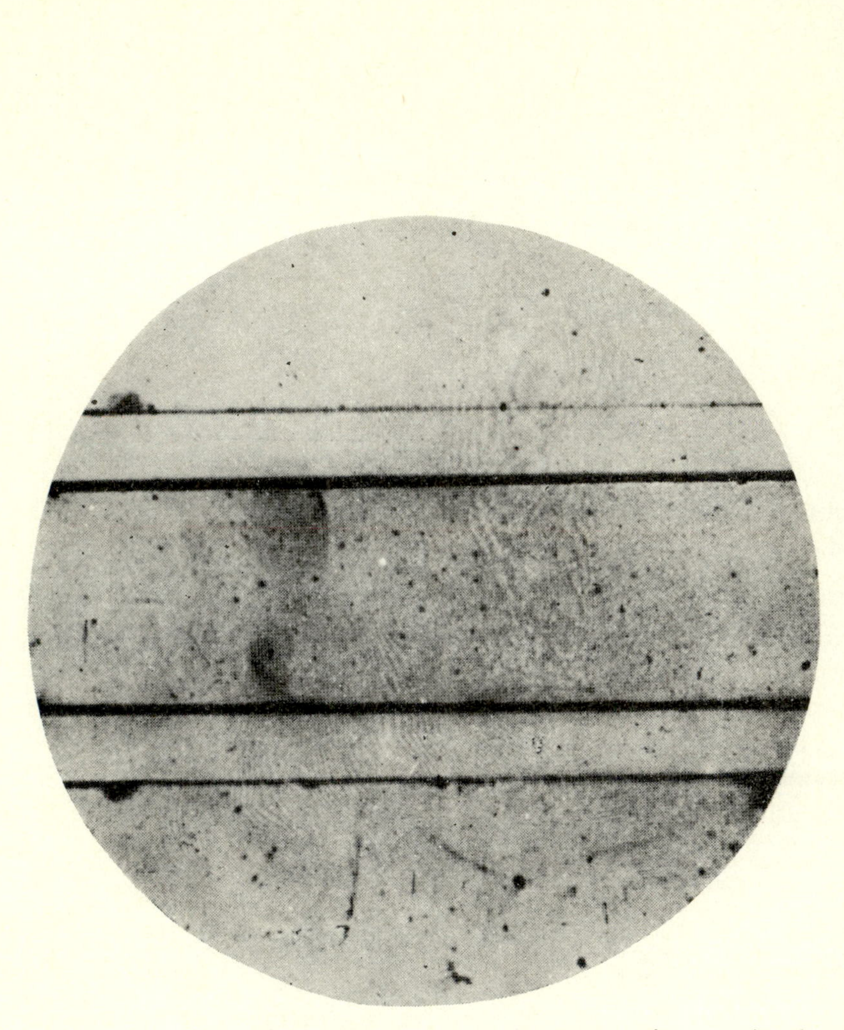

FIG. 2. A positron of 20 million volts energy ($H\varrho = 7{\cdot}1 \times 10^4$ gauss-cm) and a negatron of 30 million volts energy ($H\varrho = 10{\cdot}2 \times 10^4$ gauss-cm) projected from a plate of lead. The range of the positive particle precludes the possibility of ascribing it to a proton of the observed curvature.

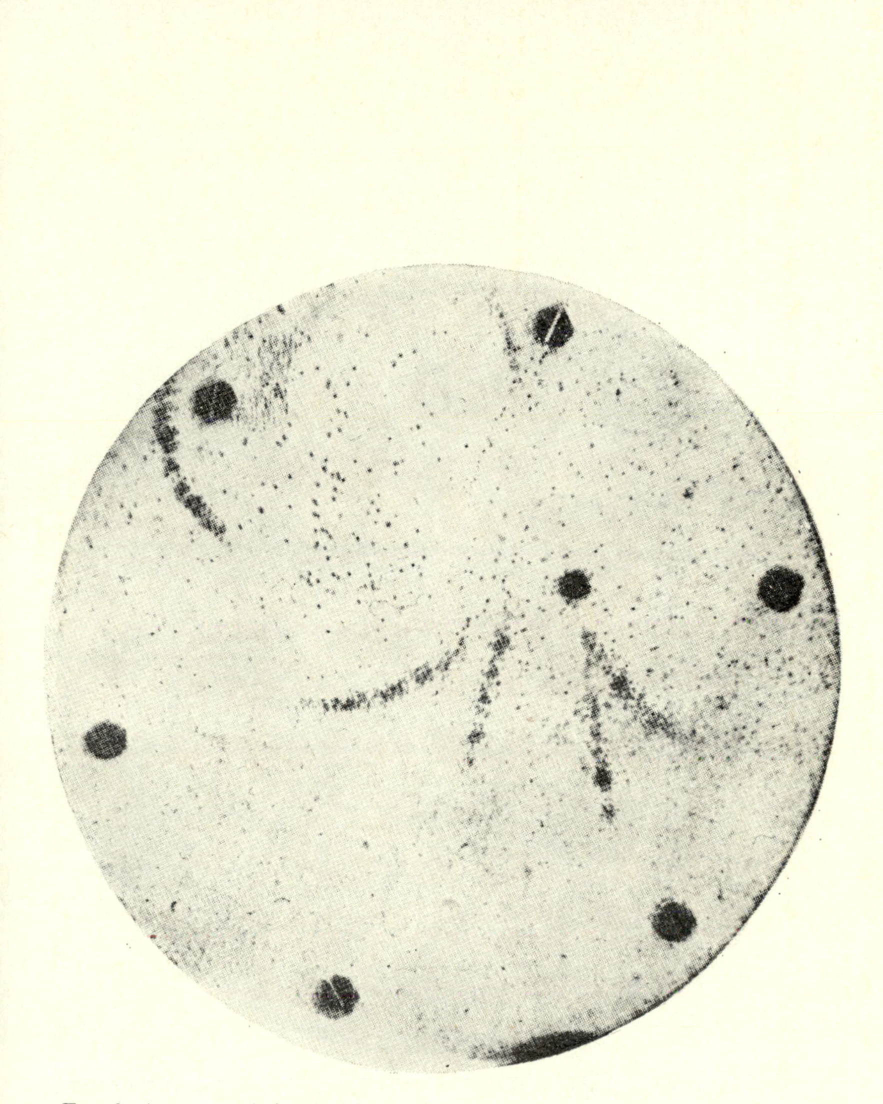

FIG. 3. A group of six particles projected from a region in the wall of the chamber. The track at the left of the central group of four tracks is a negatron of about 18 million volts energy ($H\varrho = 6{\cdot}2\times10^4$ gauss-cm) and that at the right a positron of about 20 million volts energy ($H\varrho = 7{\cdot}0\times10^4$ gauss-cm). Identification of the two tracks in the centre is not possible. A negatron of about 15 million volts is shown at the left. This group represents early tracks which were broadened by the diffusion of the ions. The uniformity of this broadening for all the tracks shows that the particles entered the chamber at the same time.

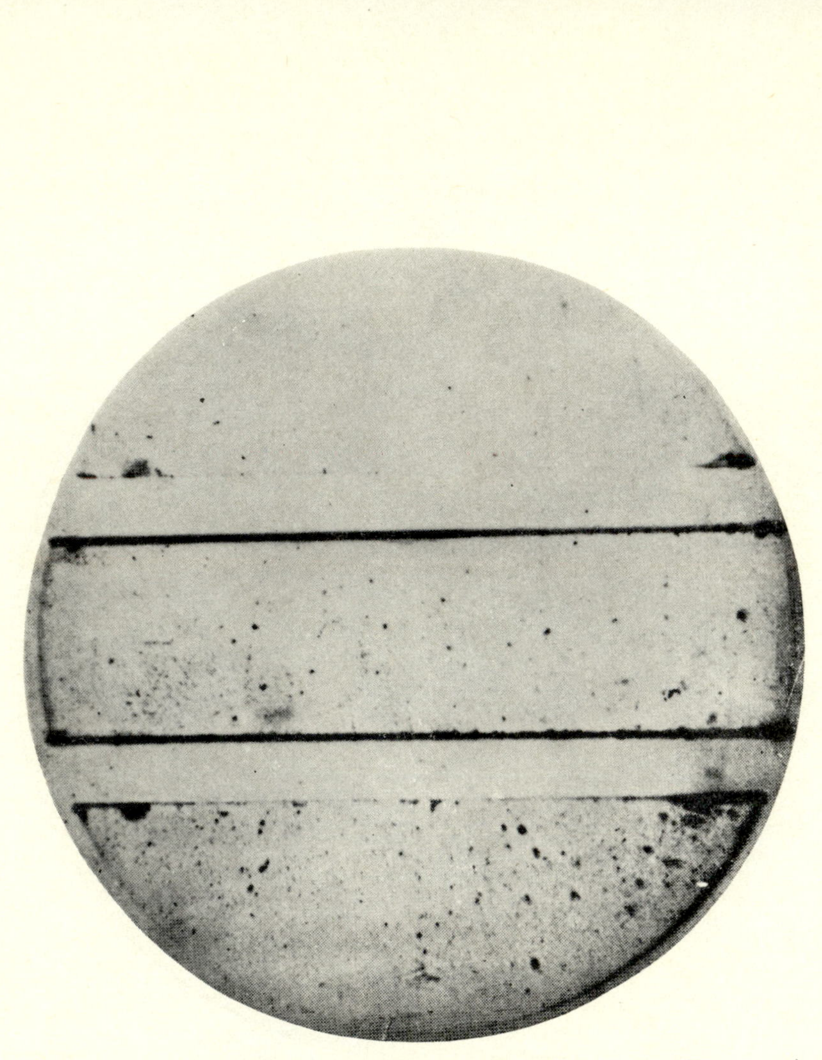

Fig. 4. A positron of about 200 million volts energy ($H\varrho = 6{\cdot}6\times10^5$ gauss-cm) penetrates the 11 mm lead plate and emerges with about 125 million volts energy ($H\varrho = 4{\cdot}2\times10^5$ gauss-cm). The assumption that the tracks represent a proton traversing the lead plate is inconsistent with the observed curvatures. The energies would then be, respectively, about 20 million and 8 million volts above and below the lead, energies too low to permit the proton to have a range sufficient to penetrate a plate of lead of 11 mm thickness.

rather than a proton and negatron in close combination, the above hypotheses will have to be abandoned for the proton will then in all probability be represented as a complex particle consisting of a neutron and positron.

While this paper was in preparation press reports have announced that P. M. S. Blackett and G. Occhialini in an extensive study of cosmic-ray tracks have also obtained evidence for the existence of light positive particles confirming our earlier report.

I wish to express my great indebtedness to Professor R. A. Millikan for suggesting this research and for many helpful discussions during its progress. The able assistance of Mr. Seth H. Neddermeyer is also appreciated.

5. On Multiplicative Showers†

J. F. Carlson and J. R. Oppenheimer‡

In I we discuss the status of the quantum theoretic formulae for pair production and radiation in the domain of cosmic-ray energies, and the relevance of these processes to an understanding of showers and bursts. In II we give a qualitative estimate of the course implied by the theory for a shower or burst built up by multiplication from a very energetic primary; we then set up the diffusion equations for the equilibrium of electrons and gamma-rays, and show how these can be simplified. In III we carry through the analytic solution of the diffusion equations, and find the distribution of electrons and gamma-rays as a function of their energy, the primary energy, and the thickness and atomic number of the matter traversed. We treat the effect of ionization losses on the shower, calculate the amount of radiation of low energy to be expected, and treat transition effects in passing from one substance to another. In IV we discuss the results of the calculations, and give a summary of the conclusions to which they lead, and the difficulties.

I

In nuclear fields, gamma-rays produce pairs, and electrons lose energy by radiation. The formulae which have been deduced[1] from the quantum theory give for the probability of these processes values which, for sufficiently high energies, no longer depend upon the energy of the radiation. Because of this, the secondaries, produced by a photon or electron of very high energy, will be nearly as penetrating as the primary, so that the primary energy will soon be divided over a large

† *Phys. Rev.* **51**, 220–231 (1937). (Received December 8th, 1936.) Only the discussion, from the outer sections, is reprinted here. Section III, containing the solution of the equations, is almost entirely omitted.

‡ University of California, Berkeley, California.

[1] An account of the results and of the theory on which they are based may be found in Heitler's book, *The Quantum Theory of Radiation* (Oxford, 1936).

number of photons and electrons. It is this development and absorption of showers which we wish to investigate.

The finite limiting cross-sections for radiative loss and for pair production essentially limit the penetrating power of electrons and photons; as we shall see, 20 cm of Pb should absorb practically all such radiation if the primary energies are $< 10^5$ MeV. From this one can conclude, either that the theoretical estimates of the probability of these processes are inapplicable in the domain of cosmic-ray energies, or that the actual penetration of these rays has to be ascribed to the presence of a component other than electrons and photons. The second alternative is necessarily radical; for cloud chamber and counter experiments show that particles with the same charge as the negative electron belong to the penetrating component of the radiation; and if these are not electrons, they are particles not previously known to physics.[2]

Direct evidence for the approximate validity of the theoretical formulae is provided by the latest studies of Anderson and Neddermeyer[3] on the energy loss and pair production of electrons of energy up to 400 MeV. This evidence is still incomplete; yet it affords absolutely no indication of a breakdown of the theoretical formulae. Since there is good evidence from the altitude and latitude curves of cosmic-ray ionization, as well as from the transition curves for showers and bursts, of a component in the cosmic rays which is strongly absorbed and yet has a very high energy, it seems of interest to investigate in detail the consequences which the theoretical formulae imply for the degradation, multiplication and absorption of such radiation. We shall find in this way a model for the building up and absorption of large showers and bursts which in many important respects agrees with what is found experimentally.[4] From this we should like to derive on the one hand a further argument for the qualitative validity of the theoretical formulae,

[2] Thus Williams first suggested that penetrating cosmic rays were protons, positive and negative: E. J. Williams, *Phys. Rev.* **45**, 729 (1934). Cloud chamber evidence would favour a particle of smaller mass.

[3] C. D. Anderson and S. Neddermeyer, *Phys. Rev.* **50**, 263 (1936).

[4] An account of the experimental findings on showers and bursts and their transition effects may be found in Geiger's article, "Die Sekundaer Effekte der Kosmischen Ultra-strahlung", *Naturwiss.* (Leipzig, 1935).

and on the other for the often repeated suggestion that many showers are built up by a long succession of simple elementary processes, and not by the simultaneous ejection of a huge number of particles in one elementary act.

Here a certain caution is necessary. Cloud chamber observations have shown the existence of two fairly well differentiated types of shower.[5] In one of these, and by far the more common, only electrons, positrons and γ-rays appear to take part; the shower particles are, except for those of very low energy, well collimated, with transverse momenta of the order of a few million volts; the showers can often be seen to increase in passing through matter, and typically, if they are large, have no well-defined focus. It is these showers for which our calculations will give us some understanding. In the other and rarer type of shower, transverse momenta of the order of 100 MeV are common; the shower is usually not collimated at all; heavy recoil particles are frequently seen; and the total number of particles is usually small. It is natural to ascribe these showers to the interaction of heavy and light particles, and to accept the arguments which Heisenberg[6] has advanced to show that for such processes the probability of ejection of several particles should be of the same order of magnitude as that of one. It would seem, however, that Heisenberg's attempt to interpret on this basis the larger showers and bursts as highly multiple elementary processes is without cogent experimental foundation; and we believe that in fact it rests on an abusive extension of the formalism of the theory of the electron neutrino field.

One can then believe in the applicability of the following analysis to cosmic-ray photons and electrons and their showers, only if he admits the presence of another component to which the analysis is not at all applicable, and of other types of elementary processes, which essentially involve the heavy particles and their coupling with electrons, and which find no place in this treatment.

[5] Anderson and Neddermeyer, see reference 3; Brode, MacPherson and Starr, *Phys. Rev.* **50,** 581 (1936).

[6] W. Heisenberg, *Zeits. f. Physik,* **101,** 533 (1936).

II

We are interested then in what happens when an electron (or positron or photon) of very high initial energy passes through matter. We shall consider only the three elementary processes of pair production by photons, radiation by electrons, and ionization losses by electrons, the first two because they dominate the multiplication which makes the shower, the third because it limits the size of the shower and absorbs it. We shall not consider relatively rare multiplicative processes, such as the production by electrons of high energy electronic secondaries, the direct production of pairs by electrons, or the Compton effect. We shall suppose that all high energy particles come off forward, and neglect the angular divergence of the shower. We shall not try to treat in detail radiation of such low energy that these simplifications are invalid.

To define the probabilities of the elementary processes, we shall use simplified formulae which approximate closely to the limiting forms which the theory gives for high energies.[1]

Thus for the probability that an electron (or positron) radiate an energy in the range E, $E+\Delta E$ in passing through a thickness Δx of matter of atomic number Z, nuclear density N, we shall take

$$P\,\Delta x\,\Delta E = \frac{K\,\Delta x\,\Delta E}{E} \quad \text{with} \quad K = \frac{4Z^2e^6N}{\hbar m^2C^5}\ln\frac{200}{Z^{\frac{1}{3}}}. \tag{1}$$

For the probability that a gamma-ray of energy E make a pair of energy E', $E-E'$, we take

$$P'\,\Delta x\,\Delta E' = (K'\,\Delta x\,\Delta E')/E; \qquad E' < E; \qquad K'/K = \sigma \sim \tfrac{2}{3}. \tag{2}$$

It is convenient to measure length in terms of a variable $t = xK$. For Pb the unit of length is $\sim\frac{1}{2}$ cm, for water it is about 0·4 m. For elements as heavy as Pb, the constants K and K' can hardly be regarded as known within 20%. Formula (1) is a good approximation as long as the electronic energy is > 20 MeV: formula (2) begins to give too large results for $E < 50$ MeV, but is off by only a factor $\frac{3}{5}$ at $E = 25$ MeV. As for ionization losses, we shall suppose them to be independent of energy, and evaluate them for those energies, where their effect will be important. Thus we shall write for electrons

$$\frac{\partial E}{\partial t} = -\beta, \quad \beta = \frac{4\pi NZe^4}{KmC^2} \ln \frac{\beta}{ZRh}\,. \tag{3}$$

For Pb, β has the value 6·5 MeV. It is quite closely inversely proportional to Z.

For primary energies which are not too high, one can carry through the calculations step by step, finding how many gamma-rays are produced by the primary, how many of these are absorbed by pair production, how many gamma-rays these in turn produce. This procedure was used in computing the number of pairs theoretically to be expected in the electron traversals studied by Anderson and Neddermeyer; it has been used by Nordheim, and by Heitler and Bhabha,[7] though here with neglect of ionization losses. It clearly becomes prohibitively laborious for thicknesses and energies of the order of those involved in large showers and in bursts; and we shall try instead to solve the diffusion equations implied by (1), (2) and (3), and so obtain an insight into the course of the shower for large thicknesses and high primary energies. An extremely rough indication of what we may expect to find we can see from a quite simple argument.

Both radiation and pair production give two rays for one in the shower. Each process of this kind has about an even chance of happening in a distance $t = 1$ (somewhat less than even for pair production, somewhat more for γ-radiation). Thus the order of magnitude of the total number of electrons and γ-rays to be expected at a thickness t is 2^t. The order of magnitude of the energy loss in thickness Δt is thus $\sim \frac{1}{2}\beta 2^t \Delta t$. When the integral of this is equal to the primary energy E_0, the shower will be absorbed. Thus if T is the distance to which the

[7] We are indebted to Dr. Nordheim for writing to us of his results. See also Heitler and Bhabha, *Nature*, **138**, 401 (1936). We are further indebted to Heitler and Bhabha for sending us a manuscript of the paper in which they have extended these calculations. Their results differ from ours primarily because of their neglect of ionization losses; apart from this the agreement between their values and ours is excellent. We do not agree with their conclusion that these calculations make it possible to ascribe the greater part of sea-level cosmic radiation to degraded electrons and photons of high initial energy.

shower penetrates

$$\beta 2^T \sim 2E_0 \ln 2. \tag{4}$$

From this it follows that T will increase logarithmically with E_0 and that it will decrease slowly with decreasing Z; that the number of particles in the shower will increase with E_0 roughly linearly, and will be roughly proportional to Z. For showers of about 30 particles we should expect the maximum in the shower to come at $T \sim -\log_2 30 \sim 5$, or a little over two cm of Pb or about 7 cm of Fe. These are in fact[4] of the order of the distances at which maxima are found for the transition curves for relatively small showers. For bursts of 1000 particles the transition maxima are found[5] at about twice as great thicknesses, as our rough estimate would suggest.

To set up our diffusion equations, let us write $\gamma(t, E)\Delta E$ for the probable number of gamma-rays to be found at a thickness t in the energy range $E, E+\Delta E$ and $\mathcal{P}(t, E)\Delta E$ for the corresponding number of electrons and positrons. Then

$$\frac{\partial \gamma}{\partial t} = -\sigma\gamma + \frac{1}{E}\int_E^{E_0} \mathcal{P}(t, \varepsilon)\, d\varepsilon, \tag{5}$$

$$\frac{\partial \mathcal{P}}{\partial t} = 2\sigma \int_E^{E_0} \frac{\gamma(t, \varepsilon)}{\varepsilon}\, d\varepsilon + \beta \frac{\partial \mathcal{P}}{\partial \varepsilon} + \int_{[E]}^{E_0} \frac{\mathcal{P}(t, \varepsilon)}{\varepsilon - E}\, d\varepsilon - \mathcal{P}(t, E)\int_{[0]}^{E} \frac{d\varepsilon}{\varepsilon}. \tag{6}$$

The last two terms in (6) can be combined to give the clearly finite result

$$R = \lim_{\delta \to 0}\left\{\int_{E+\delta}^{E_0} \frac{\mathcal{P}(t, \varepsilon)}{\varepsilon - E}\, d\varepsilon - \mathcal{P}(t, E)\int_{\delta}^{E} \frac{d\varepsilon}{\varepsilon}\right\}$$

$$= -\mathcal{P}(t, E) \ln \frac{E}{E_0 - E} - \int_E^{E_0} \frac{\partial \mathcal{P}}{\partial \varepsilon}(t, \varepsilon) \ln\left(\frac{\varepsilon}{E} - 1\right) d\varepsilon \tag{7}$$

and give the change in pair distribution which comes directly from the radiative losses of the charged particles. The term in β gives the corresponding change from ionization losses, and the first term on the right in (6) gives the pair production by γ-rays. In (5) the first term gives the absorption of γ-rays by pair production, and the second term their replenishment by radiation. These equations are to be solved for the boundary conditions

$$\gamma(t=0, E) = 0; \qquad \mathcal{P}(t=0, E) = \Delta(E, E_0), \tag{8}$$

where Δ vanishes except when E is in the immediate neighborhood of E_0, and has the integral 1. It will be convenient, however, to take $\Delta(E_0, E_0) = 0$, and to choose for Δ, not a delta function, but

$$\Delta(E, E_0) = \lim_{\alpha \to \infty} \frac{\alpha^n}{E} (\ln E_0/E)^{-1} \left(\frac{E}{E_0}\right)^{\alpha} \frac{1}{\Gamma(n)}, \qquad n \geqslant 4. \tag{9}$$

The reasons for this choice will be apparent as we develop the solution.

From (5) we may write

$$\mathcal{P}(t, E) = e^{-\sigma t}(\partial^2 z)/(\partial t\, \partial E) \quad \text{with} \quad \gamma = -e^{-\sigma t} z/E. \tag{10}$$

It is not possible by differentiation to reduce (6) to a differential equation, because of the occurrence of E in the integrand of (7). We have therefore tried to replace the terms (7) by others, which would lead to a readily soluble system of differential equations, and which would still give a good representation of (7). As we whall see, we can obtain a solution in terms of integrals of elementary functions provided we write in place of (7)

$$R \to a \int_E^{E_0} \frac{\mathcal{P}(t, \varepsilon)}{\varepsilon} d\varepsilon + b\mathcal{P}(t, E) + cE \frac{\partial \mathcal{P}}{\partial E}(t, E) + dE^2 \frac{\partial^2 \mathcal{P}}{\partial E^2} + \ldots \tag{11}$$

if a, b, c, d are constants. Two elementary conditions now limit their choice.

1. The number of particles leaving a given energy range by radiation must be equal to the number entering some other range....

2. The energy lost in x must be equal to that appearing in the γ-rays.

* * * * *

$$R = 2 \int_E^{E_0} \frac{\mathcal{P}(t, \varepsilon)}{\varepsilon} d\varepsilon - 2\mathcal{P}(t, E). \tag{11a}$$

The use of this form for R corresponds to assuming a uniform distribution of energy losses instead of the $1/E$ law given by (1), and renormalizing to give the correct total energy loss. The production of γ-rays is, however, treated correctly in accordance with (1); it is only the effect of these losses on the redistribution of the electrons themselves that is falsified. Since it is difficult to give an *a priori* estimate of the error thus introduced into the solution, we have used the solutions which we have obtained to compare the values of R given by (7) and by (11a). When t is not too small, most of the particles have of course an energy $E \ll E_0$. For this case we found that the two values of R agreed within less than 5%.

* * * * *

III

* * * * *

We give in Figure 1 a plot of $N(t)$ (the total number of electrons) against t, for water, for which we have taken $\beta = 90$ MeV, and t is measured in units of $\sim 0{\cdot}4$ m. The primary energy is $\sim 2{\cdot}5 \times 10^3$ MeV. The maximum of N occurs at about 1·2 m. It would seem probable that the latitude sensitive transition effects of the cosmic-ray ionization in the upper atmosphere are to be interpreted on this basis.

* * * * *

IV

In Figure 2 we give typical energy distribution curves for the pairs, for Pb, and a primary energy of $1{\cdot}5 \times 10^5$ MeV for $t = 15, 8, 6, 4$. In the neighborhood of $\lambda = 8$ (where $\lambda = \ln E_0/E$), the ionization corrections are appreciable, but for $\lambda < 6$ the curves will hold for all substances. It will be observed that these curves follow quite closely what we would get from a distribution law k/E^2; thus except for $t = 4$ they may all be represented by a law of this form in which the exponent of E never differs from -2 by more than 5%. In fact this law gives a very good approximation to the energy distribution for $t > 0{\cdot}7\lambda$, and for all energies $E \ll E_0$ but $\gg \beta$. We have already pointed out that

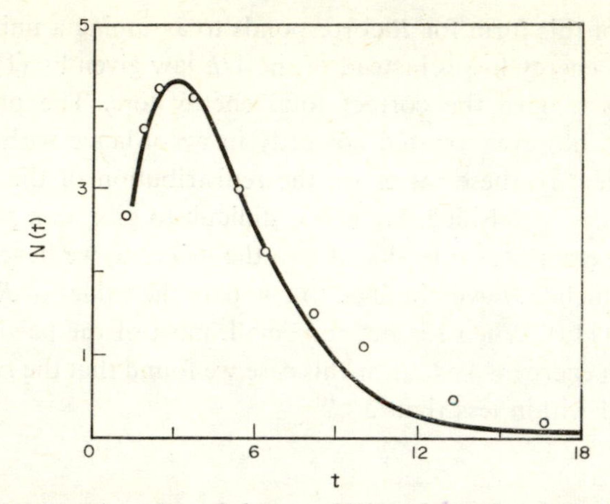

FIG. 1. Plot of total number of electrons $N(t)$ against t, for air $E_0 = 2500$ MeV computed from (36). The circles are from the experimental results of Pfotzer,† on the variations of vertical coincidence counting rate in the upper atmosphere, at the magnetic latitude 50°N. At this latitude the earth's field will just admit electrons of energy 2500 MeV. The deviations from the curve for $t > 8$ may indicate the presence of some electrons of higher energy, as well as a penetrating component.

it is for this reason that the simplification of the diffusion equations made in replacing (7) by (11) is permissible. Cloud chamber studies[8] of the energy distribution of shower electrons seem to fit this law quite well for $E > 25$ MeV, but the experimental conditions here are clearly not such as to make this agreement very significant.

Because of the form of the distribution law, a plot of $\mathcal{P}E$ will give a good estimate of the total number of electrons to be expected with energies $\geqslant E$. For Pb, $E = 50$ MeV, and several values of E_0 varying from 2700 MeV up to $1{\cdot}1 \times 10^6$ MeV we give such plots against t (Figure 3). The rapid increase with E_0 of the total number of particles, and the slow shift in the position of the maximum, confirm the qualitative arguments of § 2. We would like to interpret in terms of this shift the well-known observation that the optimum transition thickness for bursts is considerably larger than for showers. In fact, the agreement

† *Zeits. f. Physik*, **102,** 41 (1936), Fig. 1, p. 42.

[8] C. D. Anderson and S. Neddermeyer, *Int. Conf. Physics, London,* 171 (1934).

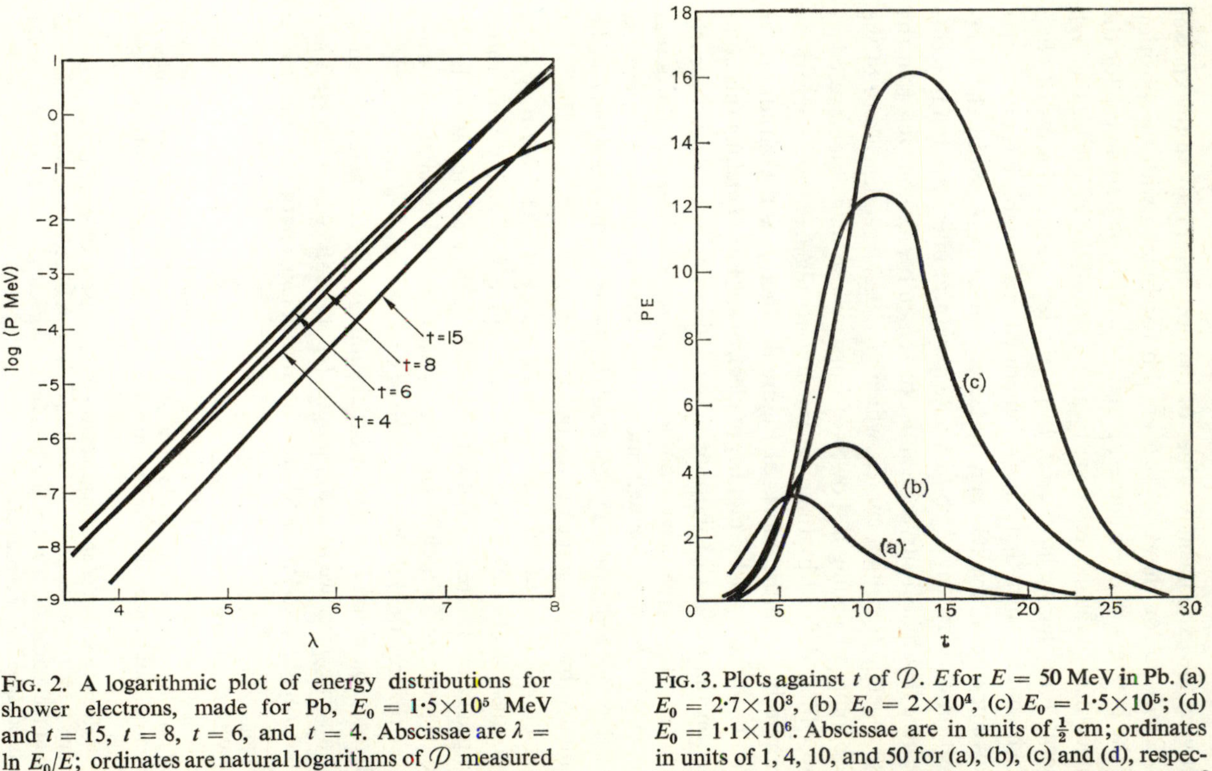

FIG. 2. A logarithmic plot of energy distributions for shower electrons, made for Pb, $E_0 = 1{\cdot}5\times10^5$ MeV and $t = 15$, $t = 8$, $t = 6$, and $t = 4$. Abscissae are $\lambda = \ln E_0/E$; ordinates are natural logarithms of $\mathcal{P}$ measured in $(\text{MeV})^{-1}$. A straight line of slope 2 would correspond to a $1/E^2$ distribution law.

FIG. 3. Plots against t of $\mathcal{P}.E$ for $E = 50$ MeV in Pb. (a) $E_0 = 2{\cdot}7\times10^3$, (b) $E_0 = 2\times10^4$, (c) $E_0 = 1{\cdot}5\times10^5$; (d) $E_0 = 1{\cdot}1\times10^6$. Abscissae are in units of $\frac{1}{2}$ cm; ordinates in units of 1, 4, 10, and 50 for (a), (b), (c) and (d), respectively. These plots also give the number of electrons of energy $\geqslant$ 50 MeV to be expected as a function of t.

between the experimental values of the optimal transition thicknesses ($\sim$ 2 cm Pb for showers, $\sim$ 5 cm Pb for bursts) with the position of the maxima of these curves, seems to us a strong argument for the correctness of the model we are treating, and of the validity of the high energy formulae we have used. In particular the experiments of Nie,[9] in which it is shown that a burst generated in a suitable layer of matter may sometimes have its magnitude very much increased by the interposition of a few cm of Pb find a very natural interpretation in terms of these theoretical curves. It must be remembered, however, that the experimental transition curves, which give the dependence on t of the probability of finding a shower of burst whose magnitude exceeds a lower limit defined by the experimental arrangement are not strictly comparable with the curves of Figure 3 which give the variation, for fixed E_0 of the probable number of electrons.[10] For in making this comparison the initial distribution of the radiation over E_0 must clearly be taken into account. We want here, too, to point out that the actual behavior of the radiation will fluctuate about that given by our curves, and that the order of magnitude of the probable fluctuations can be estimated by shifting the curves by $\Delta t \sim \pm 1$.

The application of our methods to E_0 as low as 2700 MeV may seem unjustified. Here our results, however, agree reasonably with those computed by Bhabha and Heitler[11] with neglect of ionization losses and of processes of high order. For 50 MeV the inclusion of ionization losses reduces the number of electrons to be expected by 15–20%.

It may be helpful to give a brief summary of the general results. For any absorber we measure length (t) in units which are proportional, roughly, to $Z^2\varrho/A$, where Z is the nuclear charge, A the atomic

[9] Nie, *Zeits. f. Physik*, **99**, 776 (1936).

[10] When counters are used to detect showers, the angular divergence of the shower rays is necessarily exploited. If the theory here developed is at all correct, most of these rays must have a relatively low energy, of the order of several million volts. In fact only cloud chamber observation can tell us much about radiation of much higher energy.

[11] Reference 8. Heitler and Bhabha give for $E_0 = 2{\cdot}7 \times 10^3$ MeV, a value $\mathcal{P}E = 4$, maximum for $t = 4$; we get $\mathcal{P}E = 4{\cdot}5$ at $t = 4$. This discrepancy, and the fact that we find the maximum at $t \sim 5$, are both in the direction to be expected from the effects of higher order processes.

weight, and ϱ the density, and we define the characteristic energy β (see (3)), which varies about like $1/Z$. Then,

(1) The number of electrons per unit energy is about inversely proportional to the square of the energy, as long as $\beta < E \ll E_0$, $t > \frac{1}{2}\lambda$.

(2) For given energy and $t > 1$, there are always more γ-rays than electrons; where the shower is near its maximum, this ratio varies from 1·5–2.

(3) For $E \gg \beta$, the distribution curves plotted against t are the same for all absorbers.

(4) The number of particles of energy greater than $E_1 > \beta$ passes through a maximum for a value of t which increases logarithmically with E_0/E_1 and is always quite close to $\ln_2 E_0/E_1$.

(5) The maximum number of particles with an energy less than some small multiple of β is attained for values of t which are slightly smaller than $\ln E_0/\beta$, and decrease slowly with decreasing Z. The total number of particles of energy in this range is about inversely proportional to β, or proportional to Z.

(6) The maximum size of the shower is limited only by E_0 with which it increases not quite linearly: thus an increase in E_0 by a factor of 100 gives an increase in shower size of about 70.

(7) If the initial energy E_0 is in an incident γ-ray, or is divided among a few electrons and gamma-rays, the course of the shower will be essentially unaltered.

(8) Passage of a shower from one material (1) to another (2) will increase the size of the shower if $Z_2 > Z_1$, decreases it if $Z_1 > Z_2$. The transition takes place in a thickness $t_2 \sim 1\frac{1}{2}$. All of these results apply only for energies E_0 above 10^3 MeV.

In this paper we have altogether neglected the question of how such high energy electrons and γ-rays can get down through the atmosphere. How serious this difficulty is we can see from (32), which tells us that for every electron of energy $\sim 2 \times 10^5$ MeV which hits the earth vertically, only 0·15 electron of energy > 550 MeV will survive at the earth's surface. This difficulty is made even sharper when we consider the form of the shower curves for great thicknesses $t > 30$, or the showers and bursts reported under very great thicknesses of absorber.

In fact, although when we go up far in the atmosphere, the showers, and still more markedly, the bursts, increase more rapidly than the total cosmic-ray ionization, below the atmosphere they do not fall off much more rapidly than this ionization. This suggests that, in addition to primary electrons and perhaps γ-rays, which are able to produce multiplicative showers directly, there is another cosmic-ray component, slowly absorbed, which is responsible for the continuation of the showers under thicknesses of absorber to which no electron or photon can itself penetrate. Some suggestions which we think relevant to the solution of this problem will be discussed in another paper.

6. Note on the Nature of Cosmic-ray Particles[†]

SETH H. NEDDERMEYER and CARL D. ANDERSON[‡]

MEASUREMENTS[1] of the energy loss of particles occurring in the cosmic-ray showers have shown that this loss is proportional to the incident energy and within the range of the measurements, up to about 400 MeV, is in approximate agreement with values calculated theoretically for electrons by Bethe and Heitler. These measurements were taken using a thin plate of lead (0·35 cm), and the observed individual losses were found to vary from an amount below experimental detection up to the whole initial energy of the particle, with a mean fractional loss of about 0·5. If these measurements are correct it is evident that in a much thicker layer of heavy material multiple losses should become much more important, and the probability of observing a particle loss less than a large fraction of its initial energy should be very small. For the purpose of testing this inference and also for checking our previous measurements[2] which had shown the presence of some particles less massive than protons but more penetrating than electrons obeying the Bethe–Heitler theory, we have taken about 6000 counter-tripped photographs with a 1 cm plate of platinum placed across the center of the cloud chamber. This plate is equivalent in electron thickness to 1·96 cm of lead, and to 1·86 cm of lead for a Z^2 absorption. The results of 55 measurements on particles in the range

† *Phys. Rev.* **51**, 884–886 (1937). (Received March 30th, 1937.)

‡ California Institute of Technology, Pasadena, California.

1 Anderson and Neddermeyer, *Phys. Rev.* **50**, 263 (1936).

2 Anderson and Neddermeyer, *Report of London Conference*, Vol. 1 (1934), p. 179.

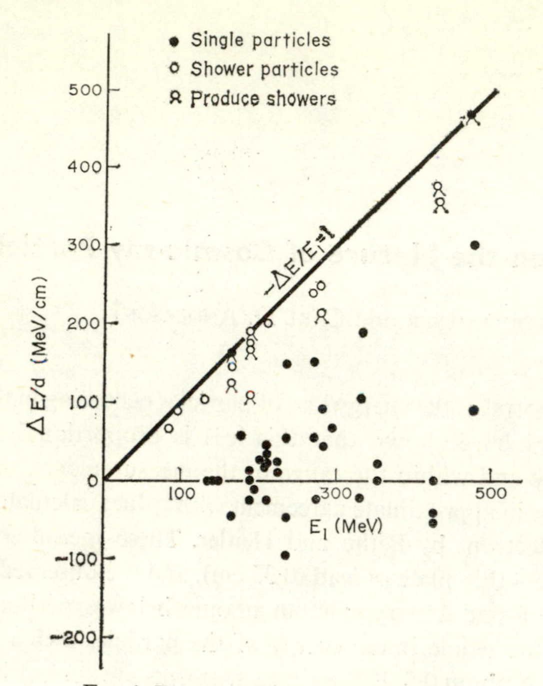

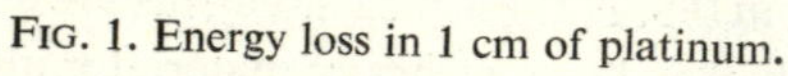
FIG. 1. Energy loss in 1 cm of platinum.

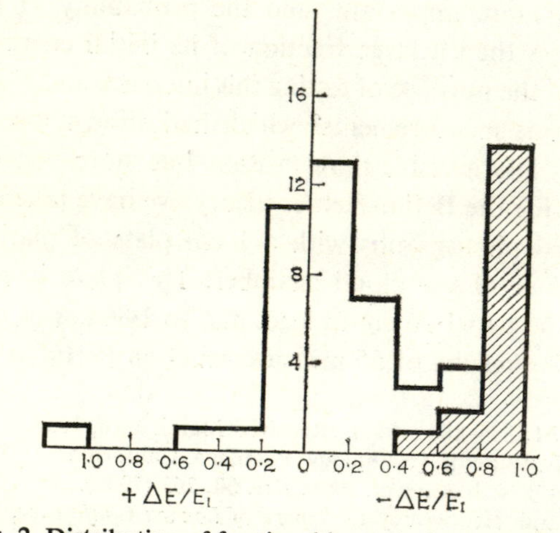

FIG. 2. Distribution of fractional losses in 1 cm of platinum.

below 500 MeV are given in Figure 1, and in Figure 2 the distribution of particles is shown as a function of the fraction of energy lost. The shaded part of the diagram represents particles which either enter the chamber accompanied by other particles or else themselves produce showers in the bar of platinum. It is clear that the particles separate themselves into two rather well-defined groups, the one consisting largely of shower particles and exhibiting a high absorbability, the other consisting of particles entering singly which in general lose a relatively small fraction of their initial energy, although there are four cases in which the loss is more than 60%. A considerable part of the spread on the negative abscissa can be accounted for by errors; it seems likely, however, that the case plotted at the extreme left represents a particle moving upward. Particles of both signs are distributed over the whole diagram, and, moreover, the initial energies of the particles of each group are distributed over the whole measured range.

The chief source of error in these experiments lies not in the curvature measurements themselves, but in the track distortions produced by irregular motions of the gas in the chamber. The distortions are much larger when a thick plate is inside the chamber than when it is left unobstructed. These distortions are not sufficient to alter essentially the distribution of observed losses for the nonpenetrating group, but they could have a very serious effect in the part of the distribution representing small losses. This is especially true inasmuch as this group represents a small percentage of the total number of tracks, selected solely on the basis that they should exhibit a measurable curvature and at the same time be free from obvious distortion. The problem of measuring small energy losses is then evidently an extremely difficult one compared to that of measuring energy distributions in an unobstructed chamber. While it is possible in many cases to distinguish a distortion as such when a magnetic field is present, it is necessary to obtain independent criteria as to the reliability of the measurements; it is not a satisfactory procedure to try to do this simply by measuring curvatures of tracks taken with no field and comparing the curvature distribution thus found with the one obtained when the field is present. Observations made with no magnetic field indicate that serious distortions occur on about 5% of the photographs, and show that they are by no means a uniform function of the orientation and position of the track in the chamber. It is therefore not possible to correct for distortion in individual cases. When large distortions do occur, however, they are likely to obey one or both of the following correlations:

(a) a curvature concave upward when the track makes a considerable angle with the vertical;

(b) a curvature concave toward the center of the chamber.

The observed percentages of measured single tracks obeying the

TABLE 1. CORRELATIONS BETWEEN TRACK CURVATURES AND POSITIONS AND ORIENTATIONS OF TRACKS

Designation (See text)	Percentage correlation		
	Observed	Expected	Observed (excluding apparent gains)
(a)	52	50	55
(b)	67	50	55
(a) and (b)	33	25	27
neither	15	25	18

correlations (a) and (b) are compared in Table 1 with the percentages expected if the observed curvatures have no relation to the positions and orientations of the tracks. If the 11 cases of apparent gains in energy are left out of consideration the observed percentages are brought somewhat closer to the expected values as shown in the last column.

A second independent check on the validity of the measurements can be obtained by measuring the scattering of the particles which show apparent curvatures, and comparing this with the scattering exhibited by those single tracks whose curvatures are just outside the range of measurability. In Figure 3 are shown the distributions of scattering angles (the angles projected on the plane of the chamber) for the measured single tracks and for single tracks with a radius of curvature, $\varrho \gtrsim 200$ cm (475 MeV). As it is scarcely conceivable that distortions could influence the scattering measurements by as much as 5°, these distributions constitute strong independent evidence that the measured

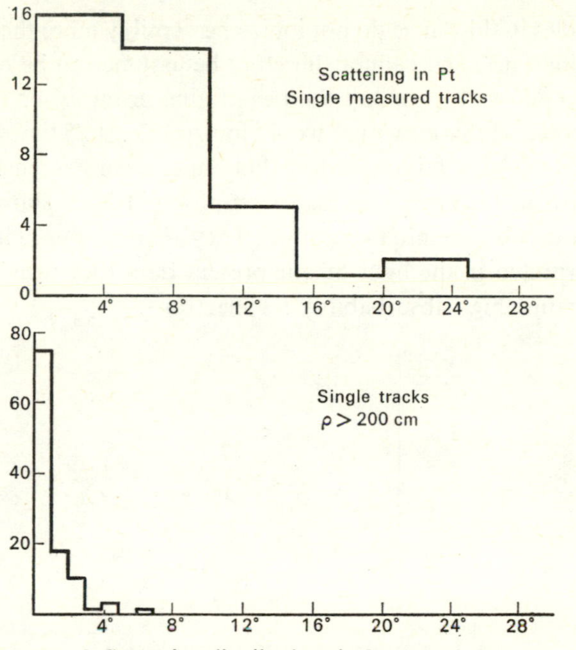

FIG. 3. Scattering distributions in 1 cm of platinum.

tracks actually lie in the energy range indicated by the curvature determinations.

It has been known for a long time that there exist particles of both penetrating and nonpenetrating types. Crussard and Leprince-Ringuet[3] have recently made measurements of energy loss in a range mainly above that covered in our experiments. They have concluded from their data that either the absorption law changes with energy or else that there is a difference in character among the particles. This same conclusion has already been twice stated by the writers.[4,5] The present data appear to constitute the first experimental evidence for the existence of particles of both penetrating and nonpenetrating character in the energy range extending below 500 MeV. Moreover, the penetrat-

[3] Crussard and Leprince-Ringuet, *C. R.* **204,** 240 (1937).

[4] Reference 2, p. 182.

[5] Reference 1, p. 268.

ing particles in this range do not ionize perceptibly more than the non-penetrating ones, and cannot therefore be assumed to be of protonic mass. The lowest $H\varrho$ among the penetrating group is $4{\cdot}5\times10^5$ gauss cm. A proton of this curvature would ionize at least 25 times as strongly as a fast electron. It is interesting that our early measurements[2] of the energy loss in thicknesses of lead from 0·7 to 1·5 cm show a similar tendency to separate into two groups. They are reproduced in Figure 4. If reinterpreted in the light of our present data they provide no evidence against high absorbability for electrons.

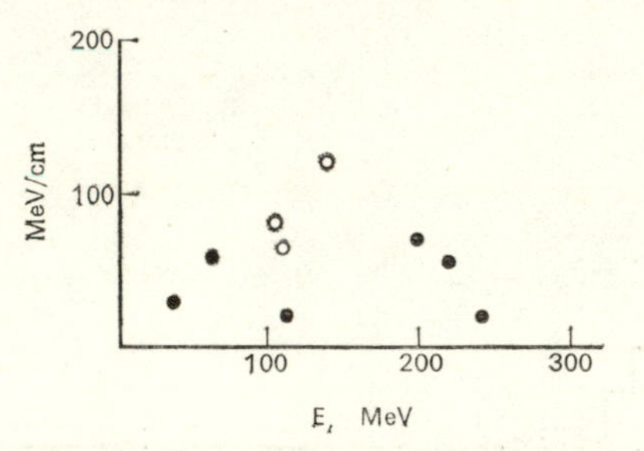

FIG. 4. Early measurements of energy loss in 0·7–1·5 cm of Pb. Dots indicate single particles; circles, shower particles.

The nonpenetrating particles are readily interpreted as free positive and negative electrons. Interpretations of the penetrating ones encounter very great difficulties, but at present appear to be limited to the following hypotheses: (a) that an electron (+ or −) can possess some property other than its charge and mass which is capable of accounting for the absence of numerous large radiative losses in a heavy element; or (b) that there exist particles of unit charge, but with a mass (which may not have a unique value) larger than that of a normal free electron[6] and much smaller than that of a proton; this assumption would also account for the absence of numerous large radiative losses, as well as for the observed ionization. Inasmuch as charge and mass are the only parameters which characterize the elec-

[6] The energies referred to throughout are, of course, calculated on the assumption of electronic mass. For a mass $m \lesssim 50m_e$ the actual energy is very roughly $E = E_{el} - mc^2$ in the range of curvature here considered.

tron in the quantum theory, assumption (b) seems to be the better working hypothesis. If the penetrating particles are to be distinguished from free electrons by a greater mass, and since no evidence for their existence in ordinary matter obtains, it seems likely that there must exist some very effective process for removing them.

The experimental fact that penetrating particles occur both with positive and negative charges suggests that they might be created in pairs by photons, and that they might be represented as higher mass states of ordinary electrons.

Independent evidence indicating the existence of particles of a new type has already been found, based on range, curvature and ionization relations; for example, Figures 12 and 13 of our previous publication.[1] In particular the strongly ionizing particle of Figure 13 cannot readily be explained except in terms of a particle of e/m greater than that of a proton. The large value of e/m apparently is not due to an e greater than the electronic charge since above the plate the particle ionizes imperceptibly differently from a fast electron, whereas below the plate its ionization definitely exceeds that of an electron of the same curvature in the magnetic field; the effects, however, are understandable on the assumption that the particle's mass is greater than that of a free electron. We should like to suggest, merely as a possibility, that the strongly ionizing particles of the type of Figure 13, although they occur predominantly with positive charge, may be related with the penetrating group above.

We wish to express our gratitude to Professor Millikan for his helpful discussions and encouragement. These experiments have been made possible by the Baker Company, who very generously loaned us the bar of platinum; and by funds supplied by the Carnegie Institution of Washington.

Note Added in Proof

Excellent experimental evidence showing the existence of particles less massive than protons, but more penetrating than electrons obeying the Bethe–Heitler theory has just been reported by Street and Stevenson, Abstract no. 40, Meeting of American Physical Society, Apr. 29, 1937.

7. Extensive Cosmic-ray Showers†

PIERRE AUGER in collaboration with P. EHRENFEST, JR., R. MAZE, J. DAUDIN, ROBLEY, and A. FREON

Introduction

It is generally admitted that the soft group present at sea level is almost entirely due to the effects of mesotrons, that is to their decay electrons and their collision electrons. These electrons should then give rise in the lower atmosphere, to local showers, the intensity of which should increase very slowly with the altitude, in the same way as the hard group (mesotrons).

But we know that the increase of the soft group with altitude is very rapid, so we must admit that electrons of another origin than that indicated above are adding their effects to those of the decay and collision electrons from mesotrons. It seems natural to suppose that they represent the end effects of the showers that the primary particles, probably electrons, which enter the high atmosphere produced there. If this is the case, we should be able to recognize it by the existence of a "coherence" of these shower particles, the multiple effects of a single initial particle remaining bound in time and in space.

We have shown the existence of these extensive showers[1] and studied their properties with counters and Wilson chambers partly at sea level, and partly in two high altitude laboratories, Jungfraujoch (3500 m) and Pic du Midi (2900 m).

† *Rev. Mod. Phys.* **11**, 288–291 (1939).

[1] P. Auger, R. Maze, T. Grivet-Meyer, *Comptes rendus*, **206**, 1721 (1938); P. Auger and R. Maze, *Comptes rendus*, **207**, 228 (1938); P. Auger, R. Maze, P. Ehrenfest, Jr., and A. Fréon, *J. de phys. et rad.* **10**, 39 (1939).

Long Distance Coincidences

If two or three counters are arranged in coincidence in free air, a small number of coincidences is observable, due to "air showers" and this number decreases quickly when the horizontal distance of the counters is increased. Schmeiser and Bothe have studied these local showers with counter separations up to half a metre.[2] If the distance is increased further, the number of coincidences decreases much more slowly, and for distances of the magnitude of ten metres, there remains a quite measurable effect, although small.[1,3] In order to continue this study efficiently, it is then necessary to use a coincidence system with a very good resolving power, especially in the use of two counters. The apparatus employed in the present work registered only multiple kicks which happened within the time lag of 10^{-6} sec.[4] The "background" or accidental coincidences can be calculated by the formula:*

$$N = \frac{n^x \tau^{(x-1)}}{60^{(x-2)}},$$

where N is the number of accidental coincidences per hour, if n is the number of single kicks in the counters per minute, x the number of counters, τ the resolving time in seconds. For instance with two large counters of 200 cm^2, we had a background of about 1 per hour, in high altitudes. With three counters, the background is always negligible.

With the simplest arrangement of two parallel and horizontal counters placed at progressively increasing distances,[5] we could obtain the results given here in logarithmic scale (Figure 1). The greatest distance was 300 m, we did not think it wise to try greater distances, because of the uncertainty of the simultaneity in the single kicks in the

[2] W. Bothe *et al.*, *Physik. Zeits.* **38**, 964 (1937).

[3] W. Kohlhörster, I. Matthes, E. Weber, *Naturwiss.* **26**, 576 (1938).

[4] R. Maze, *J. de phys. et rad.* **9**, 162 (1938).

* Instead of this equation, many writers (e.g. C. Eckart and F. R. Shonka, *Phys. Rev.* **53**, 752 (1938)) use an expression equivalent to

$$N = x\,\frac{n^x \tau^{(x-1)}}{60^{(x-2)}}$$

differing from that here given by a factor of x.

[5] P. Auger, R. Maze, and Robley, *Comptes rendus*, **208**, 1641 (1939).

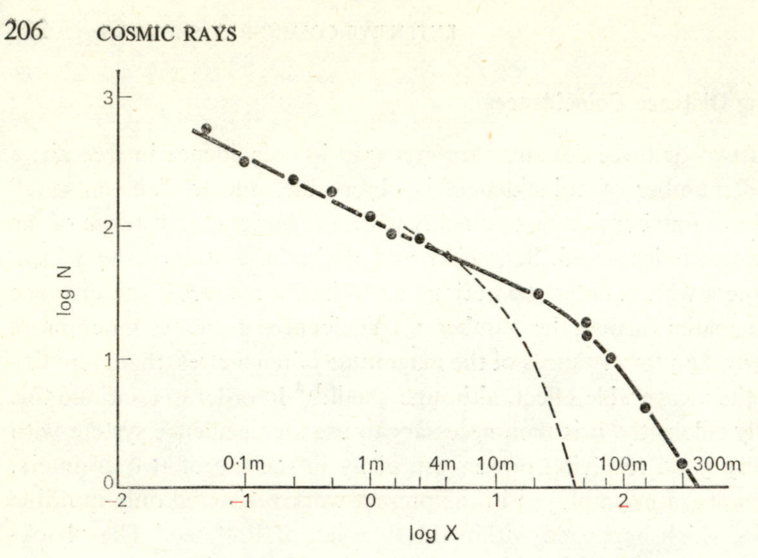

FIG. 1. Results with two parallel and horizontal counters.

counters, coming from distant particles of the same showers. The time difference can be of the same order of magnitude as the resolving power, since the showers are not necessarily vertical.

Comparison with Shower Theory

We have compared these results with the calculations made by different authors, on the basis of the cascade shower theory. Euler, for instance,[6] gives a theoretical curve for the decrease of coincidences with distance which deviates from our experimental one at about 20 m. From his calculation, the number of coincidences should be reduced to a small value at 30 m distance, while we could easily measure the coincidences at a distance ten times greater. As we have shown, these facts are in favour of a production of mesotrons in the showers, these mesotrons being able to penetrate the whole atmosphere so that with a small divergence angle they can hit the ground at large horizontal distances from the central beam of the shower, where the electrons and photons are concentrated.

[6] Private communication.

Electrons and Photons

The composition of the central beam can be proved by the study of the secondary effects of the particles in heavy material. If we cover one of the counters with lead plates of increasing thickness, we get first a strong increase in the number of coincidences (at distances less than 50 m), then, after a maximum for 1·5 cm, a decrease takes place. This shows that the particles have a shower-producing power in lead, that is to say their energy is higher than 10^7 eV. The lead plate acts as a multiplicator, and the active surface of the counter is increased. We could show that a lead plate curved in the shape of a vault of a diameter just greater than that of the counters is almost ineffective, but if the diameter is larger, the multiplication takes place, because some par-

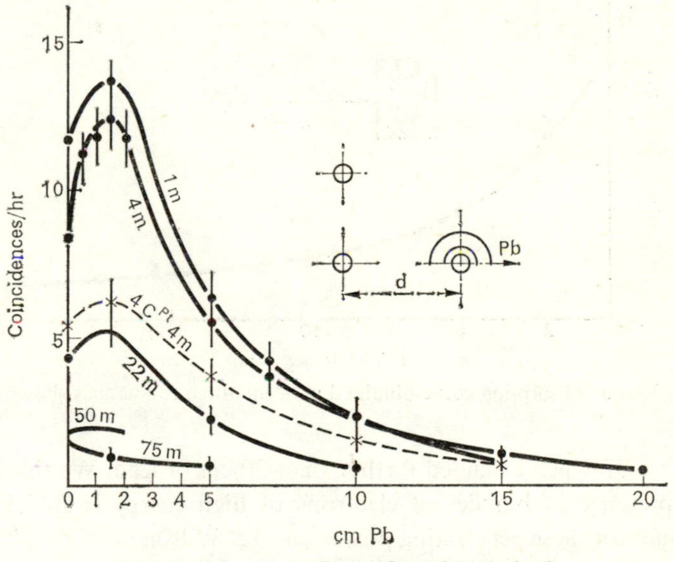

FIG. 2. Secondary effects of particles in lead.

ticles which would not have touched the counters, touch it now by the showers they produce in the lead. We could show that this multiplication effect was smaller at greater distance (Figure 2) and that consequently the particles which are responsible for the coincidences far away from the shower center are mostly electrons of smaller energy (perhaps secondaries from mesotrons?).

Absorption of Particles

We have said that thick plates reduced the number of coincidences, and this effect is due to the absorption of the particles. We tried to separate the absorption from the multiplication effect, and for that purpose we used two counters, placed above one another and which could be separated by absorbing screens, and a third one at a few metres distance. The system of two counters was protected against side showers by thick lead walls. We obtained an absorption curve

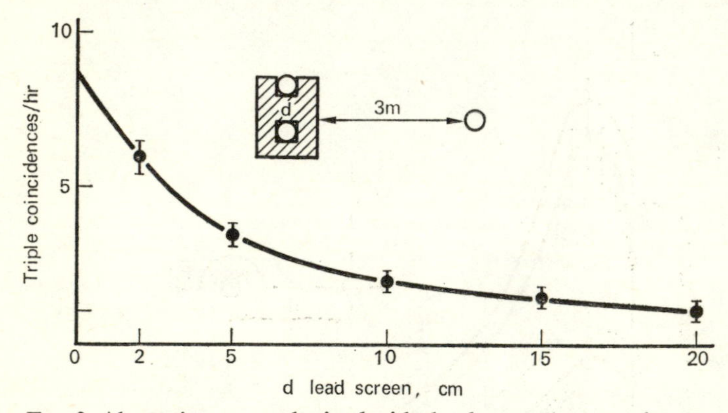

FIG. 3. Absorption curve obtained with the three counters as shown.

(Figure 3) which extended farther than 20 cm of lead. We think that the presence of bundles of electrons of high energy is sufficient to account for these penetrating powers, as our Wilson photographs have shown, up to 10 or 15 cm of lead. If the existence of a much more penetrating part is confirmed, the presence of a small proportion of mesotrons in the showers would have to be supposed.

Barometer Effect

A very strong variation of the number of showers with the barometric pressure was found. The change was as large as 15% for a variation of 15 mm Hg. The mass absorption coefficient that can be deduced from these measurements is $\mu/\varrho = 0{\cdot}007$ (in g per cm^2). From this coefficient, the decrease calculated for a screen of 3·5 m water (equivalent to 3500 m of atmosphere) is in good agreement with the ratio of showers measured in Jungfraujoch and at sea level.

Density of Tracks

The simplest method for measuring the density of tracks consists in comparing the number of coincidences with 2, 3 and 4 counters. The proportion between these numbers gives the probability for a counter to be touched by a particle in a shower. The density can be calculated, in particles per square metre, if we know the surface of the counter, and if we suppose a regular statistical distribution. We found densities from 10 to 100 particles per square metre.

But the distribution is surely not uniform, as it is for raindrops for instance. If one counter is touched, the probability of a second counter placed nearby to be touched is much greater than if the second counter is far away. This kind of distribution gives rise to local concentrations separated by relatively empty spaces. This makes the evaluation of the total number of particles a little difficult. From the measurements of density obtained at great distances (about 10 per square metre) and the total covered surface (about 10^5 square metres) we derive a total number of 10^6 particles in the large showers.

Number of Showers

With a system of three counters arranged in double coincidences we could register, at sea level, about 30 coincidences per hour. From the surface of the counters and the density of tracks we calculated that about one-half of the real showers was detected and that the number of extensive showers is of the order of 60 per hour at sea level. This number increases very rapidly with altitude; as the density of tracks changes also, the increase of the real number of extensive showers is not known with precision, but is of the order of a factor 10 between sea level and Jungfraujoch.

Wilson Chamber Photographs

We have obtained a great number of cloud-chamber photographs of portions of the extensive showers,[7] and could derive from the statistics a confirmation of the nature of the particles in the central beam; we could not prove definitely the existence of mesotrons. With a special controlling system only sensitive to the penetrating part of the showers, a number of photographs were taken, showing very dense and narrow showers. This shows that the inequalities of distribution must be taken into account in the interpretation of the absorption curves. The part of the showers which traverses 10 cm of lead seems to be constituted chiefly of narrow beams of very energetic electrons. The proportion of mesotrons in the ionizing particles is surely much smaller than 10%. An abnormally high number of slow proton tracks was also recognized on these photographs: about 12 tracks per 100 photographs, instead of 1 or 2% as we have obtained with an ordinary shower-sensitive controlling system. It shows the frequency of nuclear disintegrations in the penetrating part of the extensive showers, due perhaps to the presence of photons of very high energy and of neutrons.

Energy of Showers

A simple evaluation of the energy of these showers is obtained by the consideration of the total energy of the particles which reach the low atmosphere. We have seen that we may take 10^6 as the total number of particles present. The energy of the majority of these particles is probably the critical energy for air, that is 10^8 eV. We obtain in that way 10^{14} eV as the total energy of the shower particles, and have to

[7] See also Jánossy and Lovell, *Nature*, **142,** 716 (1938).

add a factor 10 to account for the energy lost during the traversal of the atmosphere so that 10^{15} is likely to be the energy of the primary particle.

The calculations of Bhabha–Heitler lead to the same value if we consider that the atmosphere is equivalent to 30 radiative units, and that the showers contain more than 10^5 particles. For showers of about the same number of particles reaching only the Jungfraujoch (20 radiative units) the minimum energy would be of the order of three times smaller.

Energy Spectrum of Primaries

It is interesting to see if our measurements are consistent with the spectral distribution which has been found in the case of the penetrating particles. The number N of particles with an energy higher than E is proportional to E^{-2}, so that the ratio of the numbers of showers reaching two levels should be equal to the inverse ratio of the squares of the energies of the primaries necessary to produce them. In the case of sea level ($l = 30$) and Jungfraujoch ($l = 20$), we have measured $N(20)/N(30) = 10$ and if the energies are in the ratio 1 to 3 as computed above, the agreement is good.

We may also try to give an evaluation of the number of primary particles with an energy higher than 10^{15} eV. We may reasonably think that we detect generally the showers if their densest central part falls on the counter systems, so that the effect of one incident primary extends to a surface of the order of 10^4 m^2. As we have found 60 showers per hour, it gives 10^{-9} particle of $E > 10^{15}$ eV per minute and per square centimetre in the upper atmosphere. But from the measurements in the upper atmosphere we may infer that the total number of particles with energies $E > 5 \times 10^9$ eV is of the order of 50 per minute and square centimetre. So that the ratio of numbers of these two categories of primaries is $0{\cdot}5 \times 10^{-11}$ and the inverse square ratio of their minimum energies is $2{\cdot}5 \times 10^{-11}$. The agreement is sufficient if we consider the very rough evaluations we have taken as basis.

Conclusion

One of the consequences of the extension of the energy spectrum of cosmic rays up to 10^{15} eV is that it is actually impossible to imagine a single process able to give to a particle such an energy. It seems much more likely that the charged particles which constitute the primary cosmic radiation acquire their energy along electric fields of a very great extension.

8. The Nature of the Primary Cosmic Radiation and the Origin of the Mesotron†

MARCEL SCHEIN, WILLIAM P. JESSE and E. O. WOLLAN‡

DURING the past year, our counter measurements of the vertical intensity and the production of mesotrons at high altitudes have been continued.[1] In these measurements various arrangements of counters in three-, four- or fivefold coincidences have been used. The vertical intensity has been obtained with lead thicknesses of 4, 6, 8, 10, 12 and 18 cm interposed between the counter tubes. The combined results of these experiments are plotted as curve *A* of Figure 1. It is seen from these measurements that the intensity of the hard component increases continuously to the highest altitudes reached. In our previous paper a single point of low statistical weight indicated a maximum in the mesotron curve which is not confirmed in our subsequent experiments. In the original experiments with 8 and 10 cm of lead we considered the possibility of a contribution at very high altitudes to the observed intensity from primary electrons with sufficient energy to penetrate the lead ($E > 10^{10}$ eV). This consideration led us to carry out the experiments with greater and with smaller lead thicknesses. The close agreement between the points obtained with the various thicknesses to pressures of 2 cm Hg (less than 1 radiation unit from the top of the atmosphere) is evident from the figure. This shows that measurements

† *Phys. Rev.* **59,** 615 (1941).

‡ Ryerson Physical Laboratory, University of Chicago.

[1] M. Schein, W. P. Jesse and E. O. Wollan, *Phys. Rev.* **57,** 847 (1940).

of the hard component made at very high altitudes with lead thicknesses even as small as 4 cm are not appreciably affected by electrons.

Further evidence that the traversing particles are not electrons was obtained by an arrangement of side counters registering showers generated in the lead by the traversing particles. A typical arrangement

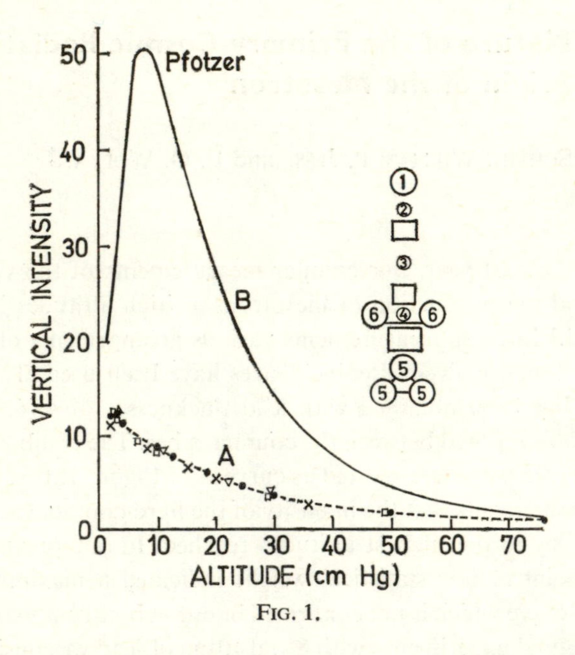

FIG. 1.

is shown in the figure in which counters 1, 2, 3, 4 and 2, 3, 4, 5 register the vertical intensities for 4 and 6 cm of lead, respectively, and counters 1, 2, 6, 4 and 2, 6, 4, 5[a] register particles accompanied by showers. If the traversing particles are electrons, there is a high probability of generating in the first 2 cm of lead a shower of many particles. In no case were more than a few percent of the traversing particles accompanied by shower counts in the side counters.

Because of the constancy of the penetrating power of the particles which we measure, and because they are not shower producing, we

[a] Superscript letters refer to Notes at end of articles.

conclude that there are no electrons of energies between 10^9 and 10^{12} eV present at the highest altitudes reached. Since the energy required for electrons to penetrate the Earth's magnetic field of 51° geomagnetic latitude is about 3×10^9 eV, and since our measurements were carried out to within the first radiation unit from the top of the atmosphere, it seems difficult to assume the presence of electrons ($E < 10^{12}$ eV) in the primary cosmic radiation and, hence, they must be replaced by some penetrating type of charged particles. The mesotrons themselves cannot be the primaries because of their spontaneous disintegration. Hence, it is probable that the incoming cosmic radiation consists of protons. The following facts support this assumption.

1. Another experiment which we have performed at high altitudes shows that mesotrons which can penetrate 18 cm of lead are produced in multiples mainly by ionizing non-shower-producing particles.[2]
2. The number of incident particles as determined by Bowen, Millikan and Neher[3] from their ionization chamber measurements at high altitudes is approximately the same as the number of penetrating particles which we observe close to the top of the atmosphere.
3. The measurements of the east–west asymmetry of cosmic rays have led Johnson[4] to suggest that the primaries of the hard component are probably protons. (In order to make it certain that all the incoming cosmic rays are positively charged, an east–west experiment for penetrating particles should be carried out at high altitudes.)

We hope to continue these observations with lower thicknesses of lead to compare with the measurements without absorption screen made by Pfotzer. Since we have assumed that the primary cosmic radiation consists of protons, the electrons known to exist in large numbers in air at high altitudes (curve *B*) must be of secondary origin. Furthermore, as seen from our experiments, the average energy of these electrons is low ($E < 10^9$ eV). These facts suggest that the elec-

[2] This process is in addition to the process already reported of the production of mesotrons by non-ionizing radiation.

[3] I. S. Bowen, R. A. Millikan and H. V. Neher, *Phys. Rev.* **53,** 217 (1938).

[4] T. H. Johnson, *Rev. Mod. Phys.* **11,** 208 (1939).

trons in the atmosphere arise mainly from the decay of the mesotron and knock-on processes.

The writers wish to express to Professor A. H. Compton their appreciation for his support of these experiments and his continued interest in them.

Note

[a]Coincident discharges of these groups of counters are implied. The counter tubes are shown in section in the diagram.

9. Observations on the Tracks of Slow Mesons in Photographic Emulsions†*

C. M. G. LATTES, G. P. S. OCCHIALINI and C. F. POWELL‡

Introduction

In recent experiments, it has been shown that charged mesons, brought to rest in photographic emulsions, sometimes lead to the production of secondary mesons. We have now extended these observations by examining plates exposed in the Bolivian Andes at a height of 5500 m, and have found, in all, forty examples of the process leading to the production of secondary mesons. In eleven of these, the secondary particle is brought to rest in the emulsion so that its range can be determined. In Part 1 of this article, the measurements made on these tracks are described, and it is shown that they provide evidence for the existence of mesons of different mass. In Part 2, we present further evidence on the production of mesons, which allows us to show that many of the observed mesons are locally generated in the "explosive" disintegration of nuclei, and to discuss the relationship of the different types of mesons observed in photographic plates to the penetrating component of the cosmic radiation investigated in experiments with Wilson chambers and counters.

† *Nature*, **160**, 453–456, and Summary from p. 492. (Received October 4th and 11th, 1947.)

* This article contains a summary of the main features of a number of lectures given, one at Manchester on June 18th and four at the Conference on Cosmic Rays and Nuclear Physics, organized by Prof. W. Heitler, at the Dublin Institute of Advanced Studies, July 5th–12th. A complete account of the observations, and of the conclusions which follow from them, will be published elsewhere.

‡ H. H. Wills Physical Laboratory, University of Bristol.

Part I. Existence of Mesons of Different Mass

As in the previous communications,[1] we refer to any particle with a mass intermediate between that of a proton and an electron as a meson. It may be emphasized that, in using this term, we do not imply that the corresponding particle necessarily has a strong interaction with nucleons, or that it is closely associated with the forces responsible for the cohesion of nuclei.

We have now observed a total of 644 meson tracks which end in the emulsion of our plates. 451 of these were found, in plates of various types, exposed at an altitude of 2800 m at the Observatory of the Pic du Midi, in the Pyrenees; and 193 in similar plates exposed at 5500 m at Chacaltaya in the Bolivian Andes. The 451 tracks in the plates exposed at an altitude of 2800 m were observed in the examination of 5 cc emulsion. This corresponds to the arrival of about 1·5 mesons per cc per day, a figure which represents a lower limit, for the tracks of some mesons may be lost through fading, and through failure to observe tracks of very short range. The true number will thus be somewhat higher. In any event, the value is of the same order of magnitude as that we should expect to observe in delayed coincidence experiments at a height of 2800 m, basing our estimates on the observations obtained in similar experiments at sea-level, and making reasonable assumptions about the increase in the number of slow mesons with altitude. It is therefore certain that the mesons we observe are a common constituent of the cosmic radiation.

Photomicrographs of two of the new examples of secondary mesons, Nos. III and IV, are shown in Figures 1 and 2. Table 1 gives details of the characteristics of all events of this type observed up to the time of writing, in which the secondary particle comes to the end of its range in the emulsion.

The distribution in range of the secondary particles is shown in Figure 3. The values refer to the lengths of the projections of the actual trajectories of the particles on a plane parallel to the surface of the emulsion. The true ranges cannot, however, be very different from the values given, for each track is inclined at only a small angle to the plane

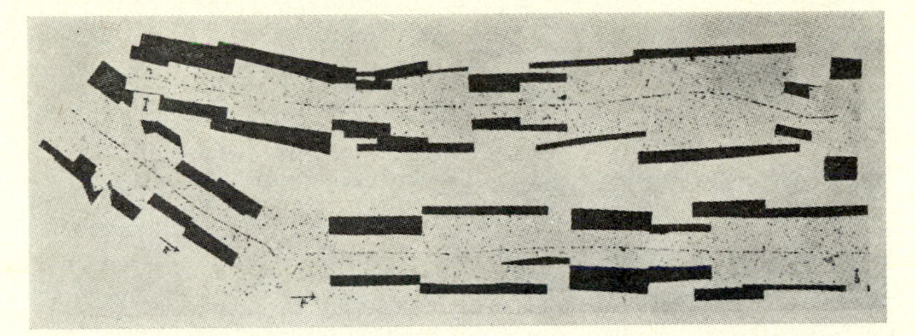

FIG. 1. Observation by Mrs. I. Powell. Cooke×95 achromatic objective; C2 Ilford Nuclear Research Emulsion loaded with boron. The track of the μ-meson is given in two parts, the point of junction being indicated by *a* and an arrow.

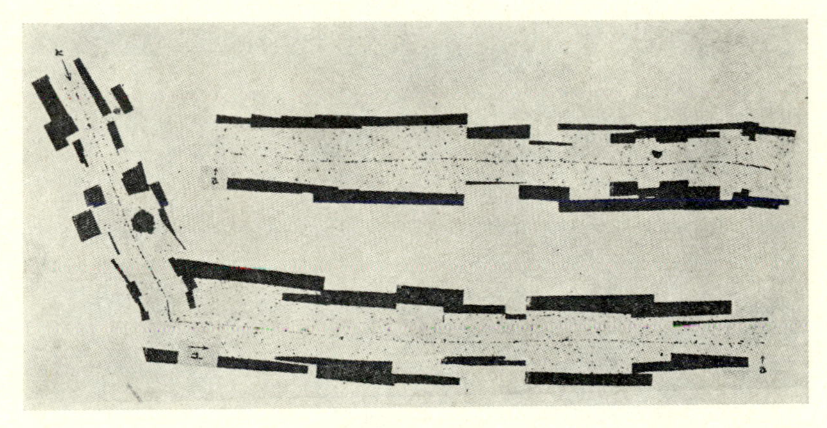

FIG. 2. Cooke×95 achromatic objective. C2 Ilford Nuclear Research Emulsion loaded with boron.

TABLE 1

Event no.	Range in emulsion in microns of	
	Primary meson	Secondary meson
I	133	613
II	84	565
III	1040	621
IV	133	591
V	117	638
VI	49	595
VII	460	616
VIII	900	610
IX	239	666
X	256	637
XI	81	590

Mean range $614 \pm 8\ \mu$. Straggling coefficient $\sqrt{(\Sigma \Delta_i^2/n)} = 4{\cdot}3\%$, where $\Delta_i = R_i - \bar{R}$, R_i being the range of a secondary meson, and $\bar{R}$ the mean value for n particles of this type.

of the emulsion over the greater part of its length. In addition to the results for the secondary mesons which stop in the emulsion, and which are represented in Figure 3 by black squares, the length of a number of tracks from the same process, which pass out of the emulsion when near the end of their range, are represented by open squares.

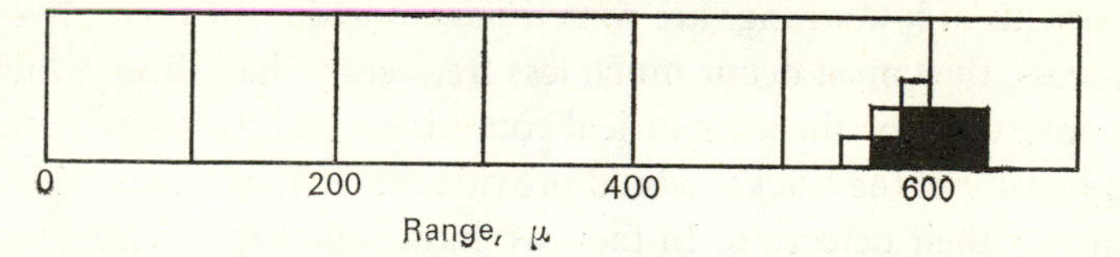

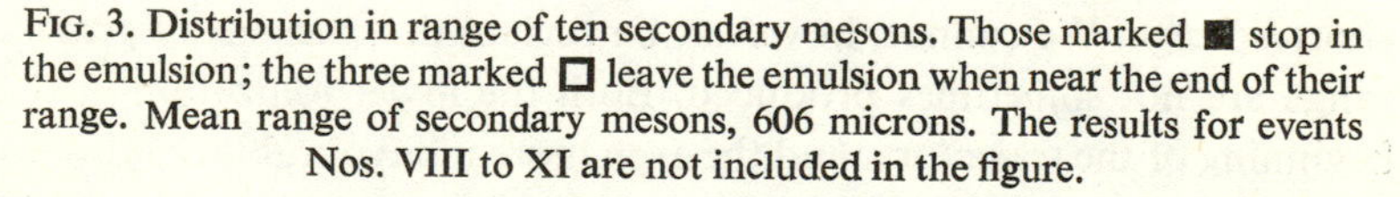
FIG. 3. Distribution in range of ten secondary mesons. Those marked ■ stop in the emulsion; the three marked □ leave the emulsion when near the end of their range. Mean range of secondary mesons, 606 microns. The results for events Nos. VIII to XI are not included in the figure.

THE μ-DECAY OF MESONS

Two important conclusions follow from these measurements. Our observations show that the directions of ejection of the secondary mesons are orientated at random. We can therefore calculate the probability that the trajectory of a secondary meson, produced in a process of the type which we observe, will remain within the emulsion, of thickness 50 μ, for a distance greater than 500 μ. If we assume, as a first approximation, that the trajectories are rectilinear, we obtain a value for the probability of 1 in 20. The marked Coulomb scattering of mesons in the Nuclear Research emulsions will, in fact, increase the probability of "escape". The six events which we observe in plates exposed at 2800 m, in which the secondary particle remains in the emulsion for a distance greater than 500 μ, therefore correspond to the occurrence in the emulsion of 120 ± 50 events of this particular type. Our observations, therefore, prove that the production of a secondary meson is a common mode of decay of a considerable fraction of those mesons which come to the end of their range in the emulsion.

Second, there is remarkable consistency between the values of the range of the secondary mesons, the variation among the individual values being similar to that to be expected from "straggling", if the particles are always ejected with the same velocity. We can therefore conclude that the secondary mesons are all of the same mass and that they are emitted with constant kinetic energy.

If mesons of lower range are sometimes emitted in an alternative type of process, they must occur much less frequently than those which we have observed; for the geometrical conditions, and the greater average grain-density in the tracks, would provide much more favourable conditions for their detection. In fact, we have found no such mesons of shorter range. We cannot, however, be certain that mesons of greater range are not sometimes produced. Both the lower ionization in the beginning of the trajectory, and the even more unfavourable conditions of detection associated with the greater lengths of the tracks, would make such a group, or groups, difficult to observe. Because of the large fraction of the mesons which, as we have seen, can be attributed to the observed process, it is reasonable to assume that alternative modes of decay, if they exist, are much less frequent than that which we have observed. There is, therefore, good evidence for the production of a single homogeneous group of secondary mesons, constant in mass and kinetic energy. This strongly suggests a fundamental process, and not one involving an interaction of a primary meson with a particular type of nucleus in the emulsion. It is convenient to refer to this process in what follows as the μ-decay. We represent the primary mesons by the symbol π, and the secondary by μ. Up to the present, we have no evidence from which to deduce the sign of the electric charge of these particles. In every case in which they have been observed to come to the end of their range in the emulsion, the particles appear to stop without entering nuclei to produce disintegrations with the emission of heavy particles.

Knowing the range-energy relation for protons in the emulsion, the energy of ejection of the secondary mesons can be deduced from their observed range, if a value of the mass of the particles is assumed. The values thus calculated for various masses are shown in Table 2.

TABLE 2

Mass in m_e	100	150	200	250	300
Energy in MeV	3·0	3·6	4·1	4·5	4·85

No established range-energy relation is available for protons of energies above 13 MeV, and it has therefore been necessary to rely on an extrapolation of the relation established for low energies. We estimate that the energies given in Table 2 are correct to within 10%.

EVIDENCE OF A DIFFERENCE IN MASS OF π- AND μ-MESONS

It has been pointed out[1] that it is difficult to account for the μ-decay in terms of an interaction of the primary meson with the nucleus of an atom in the emulsion leading to the production of an energetic meson of the same mass as the first. It was therefore suggested that the observa-

[1] *Nature*, **159**, 93, 186, 694 (1947).

tions indicate the existence of mesons of different mass. Since the argument in support of this view relied entirely on the principle of the conservation of energy, a search was made for processes which were capable of yielding the necessary release of energy, irrespective of their plausibility on other grounds. Dr. F. C. Frank has re-examined such possibilities in much more detail, and his conclusions are given in an article to follow. His analysis shows that it is very difficult to account for our observations, either in terms of a nuclear disintegration, or of a "building-up" process in which, through an assumed combination of a negative meson with a hydrogen nucleus, protons are enabled to enter stable nuclei of the light elements with the release of binding energy. We have now found it possible to reinforce this general argument for the existence of mesons of different mass with evidence based on grain-counts.

We have emphasized repeatedly[2] that it is necessary to observe great caution in drawing conclusions about the mass of particles from grain-counts. The main source of error in such determinations arises from the fugitive nature of the latent image produced in the silver halide granules by the passage of fast particles. In the case of the μ-decay process, however, an important simplification occurs. It is reasonable to assume that the two meson tracks are formed in quick succession, and are subject to the same degree of fading. Secondly, the complete double track in such an event is contained in a very small volume of the emulsion, and the processing conditions are therefore identical for both tracks, apart from the variation of the degree of development with depth. These features ensure that we are provided with very favourable conditions in which to determine the ratio of the masses of the π- and μ-mesons, in some of these events.

In determining the grain density in a track, we count the number of individual grains in successive intervals of length 50 μ along the trajectory, the observation being made with optical equipment giving large magnification ($\times$2000), and the highest available resolving power. Typical results for protons and mesons are shown in Figure 4. These results were obtained from observations on the tracks in a single plate, and it will be seen that there is satisfactory resolution between the

[2] See ref. 1 on previous page.

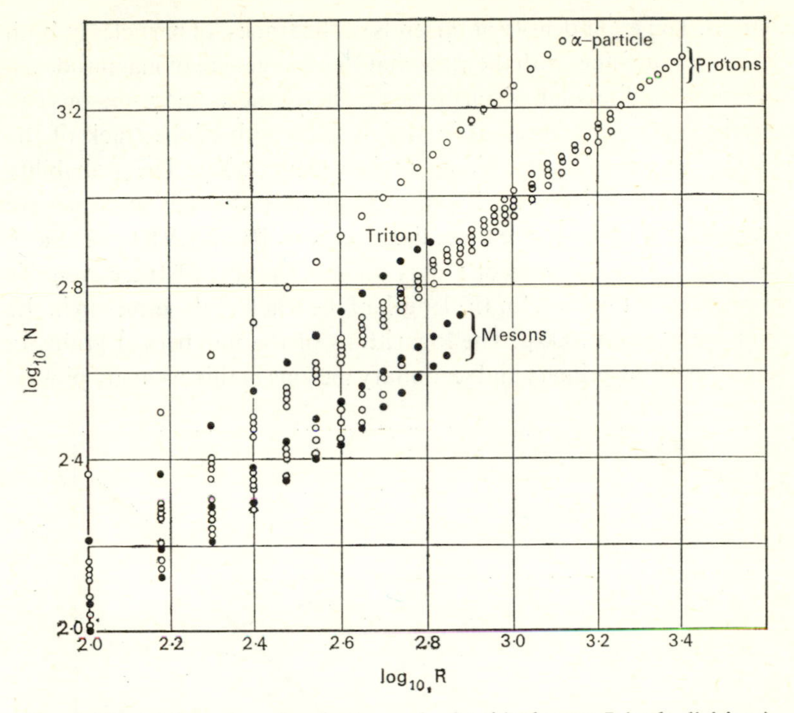

FIG. 4. N is total number of grains in track of residual range R (scale-divisions). 1 scale-division = 0·85 microns.

curves for particles of different types. The "spread" in the results for different particles of the same type can be attributed to the different degrees of fading associated with the different times of passage of the particles through the emulsion during an exposure of six weeks.

Applying these methods to the examples of the μ-decay process, in which the secondary mesons come to the end of their range in the emulsion, it is found that in every case the line representing the observations on the primary meson lies above that for the secondary particle. We can therefore conclude that there is a significant difference in the grain-density in the tracks of the primary and secondary mesons, and therefore a difference in the mass of the particles. This conclusion depends, of course, on the assumption that the π- and μ-particles carry equal charges. The grain-density at the ends of the tracks, of particles of both types, are consistent with the view that the charges are of magnitude $|e|$.

A more precise comparison of the masses of the π- and μ-mesons can only be made in those cases in which the length of the track of the primary meson in the emulsion is of the order of 600 μ. The probability of such a favourable event is rather small, and the only examples we have hitherto observed are those listed as Nos. III and VIII in Table 1. A mosaic of micrographs of a part only of the first of these events is reproduced in Figure 1, for the length of the track of the μ-meson in the emulsion exceeds 1000 μ. The logarithms of the numbers of grains in the tracks of the primary and secondary mesons in this event are plotted

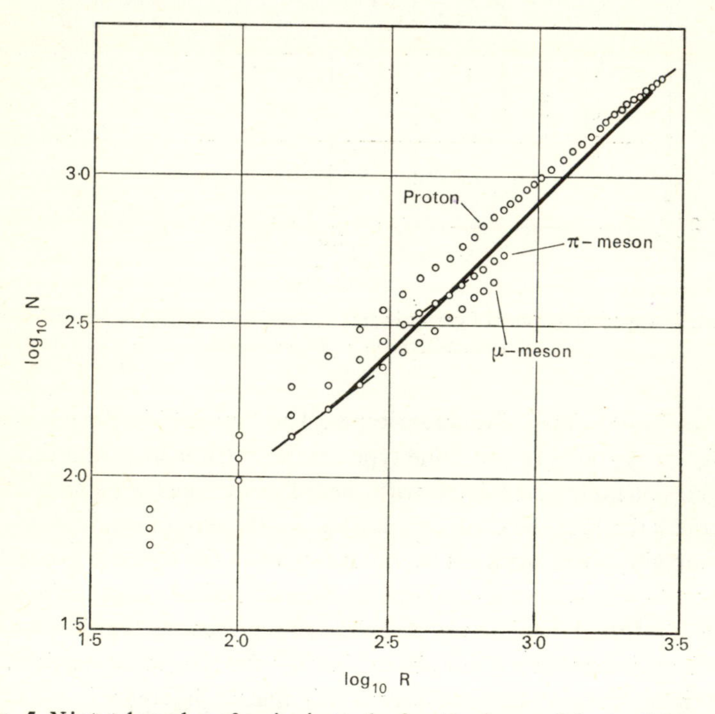

FIG. 5. N is total number of grains in track of residual range R (scale-divisions).
1 scale-division = 0·85 microns
The 45° line cuts the curves of the mesons and proton in the region of the same grain density.

against the logarithm of the residual range in Figure 5. By comparing the residual ranges at which the grain-densities in the two tracks have the same value, we can deduce the ratio of the masses. We thus obtain the result $m_\pi/m_\mu = 2{\cdot}0$. Similar measurements on event No. VIII give the value 1·8. In considering the significance which can be attached to this result, it must be noticed that in addition to the standard deviations in the number of grains counted, there are other possible sources of error. Difficulties arise, for example, from the fact that the emulsions do not consist of a completely uniform distribution of silver halide grains. "Islands" exist, in which the concentration of grains is significantly higher, or significantly lower, than the average values, the variations being much greater than those associated with random fluctuations. The measurements on the other examples of μ-decay are much less reliable on account of the restricted range of the π-mesons in the emulsion; but they give results lower than the above values. We think it unlikely, however, that the true ratio is as low as 1·5.

The above result has an important bearing on the interpretation of the μ-decay process. Let us assume that it corresponds to the spontaneous decay of the heavier π-meson, in which the momentum of the μ-meson is equal and opposite to that of an emitted photon. For any assumed value of the mass of the μ-meson, we can calculate the energy of ejection of the particle from its observed range, and thus determine its momentum. The momentum, and hence the energy of the emitted photon, is thus defined; the mass of the π-meson follows from the relation

$$c^2 m_\pi = c^2 m_\mu + E_\mu + h\nu.$$

It can thus be shown that the ratio m_π/m_μ is less than 1·45 for any assumed value of m_μ in the range from 100 to $300m_e$, m_e being the mass of the electron (see Table 3). A similar result is obtained if it is assumed that a particle of low mass, such as an electron or a neutrino, is ejected in the opposite direction to the μ-meson.

On the other hand, if it is assumed that the momentum balance in the μ-decay is obtained by the emission of a neutral particle of mass equal to the μ-meson mass, the calculated ratio is about 2·1 : 1.

TABLE 3

Assumed mass m_μ	E (MeV)	$h\nu$ (MeV)	m_π	$m_\pi/m_\mu \pm 3\%$
100 m	3·0	17	140 m_e	1·40
150	3·6	23	203	1·35
200	4·1	29	264	1·32
250	4·5	34	325	1·30
300	4·85	39	387	1·29

Our preliminary measurements appear to indicate, therefore, that the emission of the secondary meson cannot be regarded as due to a spontaneous decay of the primary particle, in which the momentum balance is provided by a photon, or by a particle of small rest-mass. On the other hand, the results are consistent with the view that a neutral particle of approximately the same rest-mass as the μ-meson is emitted. A final conclusion may become possible when further examples of the μ-decay, giving favourable conditions for grain-counts, have been discovered.

* * * * *

Summary

We may summarize our results in the following terms. We regard our observations as establishing:

(a) that two types of mesons exist, of different mass, which we refer to as π- and μ-mesons;

(b) taken in conjunction with the observations of D. H. Perkins, that some slow mesons can enter nuclei to produce disintegrations with the emission of heavy particles; we regard these mesons as negatively charged and refer to them, provisionally, as σ-mesons; and

(c) that σ-, and possibly π-mesons, can be produced in processes associated with explosive disintegration of nuclei.

Our observations suggest:

(d) that π-, and a large proportion of the σ-mesons are, respectively, positively and negatively charged particles of the same type, which interact strongly with nucleons, the negatively charged mesons being captured by both light and heavy nuclei to produce disintegrations with the emission of heavy particles;

(e) that the heavier π-mesons suffer spontaneous decay with the emission of the lighter μ-mesons; and that in the decay process the momentum of the μ-meson is balanced by that of a neutral particle of approximately the same mass.

In addition, the main features of the present experimental results, and those obtained in delayed coincidence and Wilson chamber experiments, can be explained on the assumption that:

(f) the greater part of the mesons observed at sea-level are μ-mesons formed by the decay in flight of π-mesons; and

(g) that positive and negative π-mesons are short-lived, with a mean life-time in the interval from 10^{-6} to 10^{-11} sec.

We have pleasure in acknowledging our indebtedness to Prof. W. Heitler, Prof. C. Møller, Dr. H. Fröhlich, Prof. L. Jánossy and other members of the Dublin colloquia, held in July, for a number of discussions which were of great value to us in the formulation of our ideas. Our thanks are also due to Prof. J. Baillaud and Dr. M. Hugon of l'Observatoire du Pic du Midi, and to M. René Varin, cultural attaché of the French Embassy in London, for further assistance in securing exposures; and to Prof. Ismael Escobar, director of the Bolivian Meteorological Service, for assistance during the course of the expedition to the Andes.

10. Evidence for Multiple Meson and γ-ray Production in Cosmic-ray Stars†

M. F. KAPLON, B. PETERS and H. L. BRADT‡

IN a survey of electron-sensitive Kodak NTB3 plates flown at an altitude of 100,000 ft we have observed a collision of an extremely energetic primary cosmic-ray α-particle with a heavy nucleus (Ag or Br) of the emulsion; this collision gives rise to a very narrow penetrating shower core of some 23 relativistic singly charged particles, this core being surrounded by a more diffuse shower of 33 relativistic particles (Figure 1). In addition, 18 non-relativistic[a] heavy particles, carrying at least 23 units of charge and a total energy of $\sim$ 3 BeV, emerge from the star at the origin of the shower.

The dense core (C) of the shower of relativistic particles, the majority of which are assumed to be mesons, has a total projected angular spread of 2·5°. The axis of the core is an exact continuation of the direction of the incident α-particle. The core is so dense that not all individual tracks are resolvable in the immediate vicinity of the star. At larger distances where resolution is possible all these tracks prove to be minimum ionization tracks. By counting over a given distance the number of grains in the core and dividing by the number of grains of a minimum ionization track we obtain $N_0^{(e)} = 23 \pm 2$ as estimates of the number of charged relativistic particles in the narrow shower.

The region where the narrow shower C, after passage through some 2 cm glass, penetrat es the next plate of the stack (the dotted area C of

† *Phys. Rev.* **76**, 1735–1736 (1949). (Received October 10th, 1949.)
‡ University of Rochester, Rochester, New York.
[a] Superscript letters refer to Notes at end of articles.

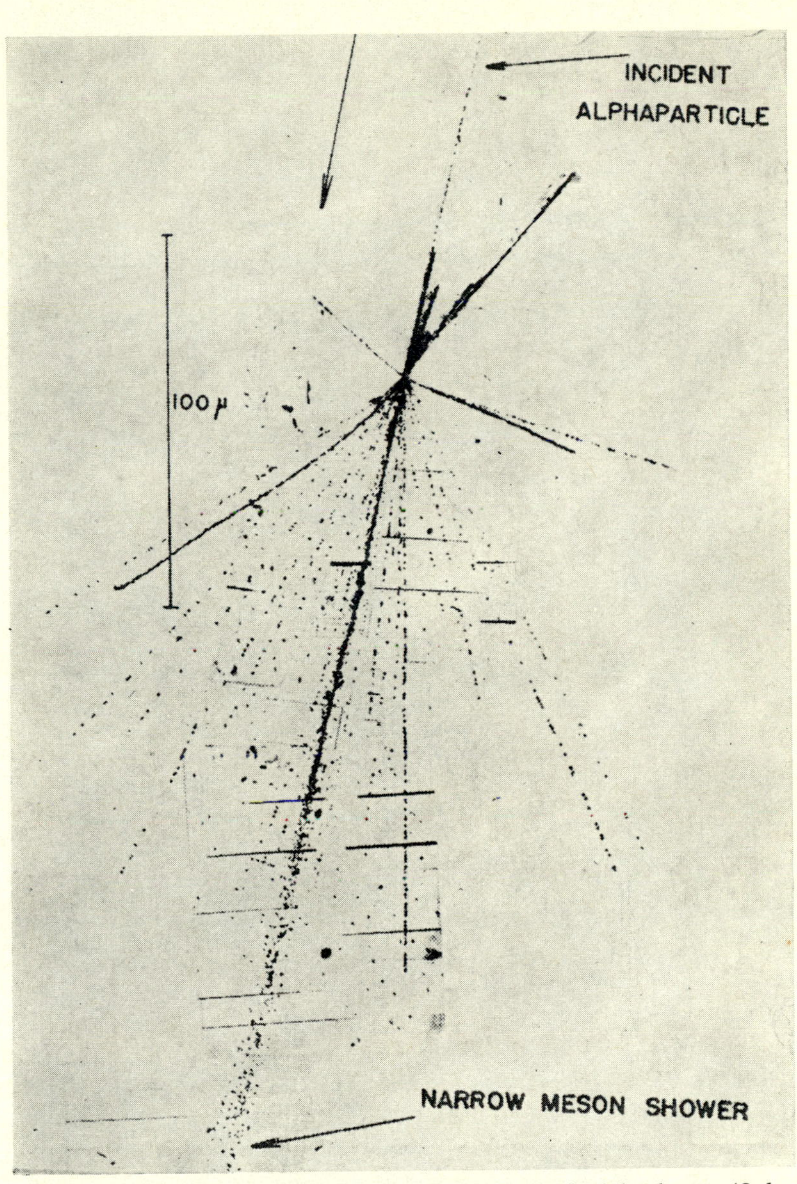

FIG. 1. Incomplete photo-micrograph (slightly retouched) of the shower. (Only a small fraction of the heavy prongs are shown.)

0·25 cm² in Figure 2) was surveyed for tracks of such length and orientation that they can be assumed to be caused by the explosion; 44 tracks of relativistic particles were found, 38 inside and six near the edge of the dotted cone C, that is 11 more tracks than were originally contained in

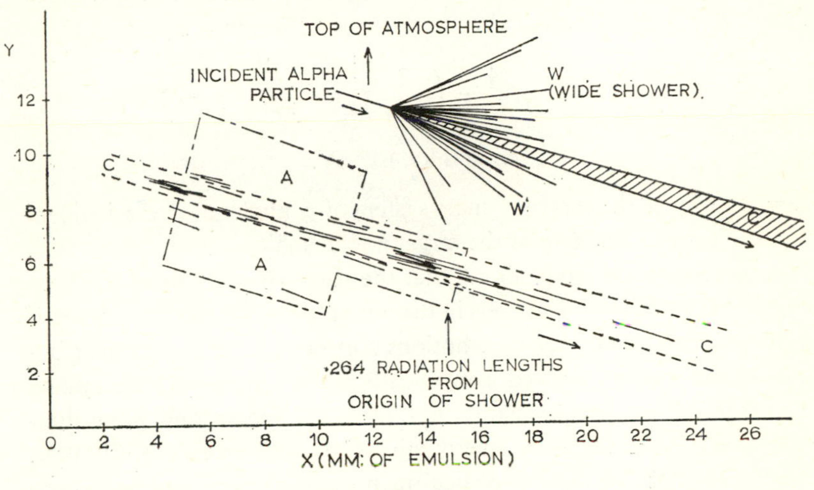

FIG. 2.

the narrow shower. In surveying an additional area $A = 0{\cdot}29$ cm² for tracks satisfying identical criteria only six more tracks were found, all but one lying very close to the core and probably also belonging to the shower. We conclude that considerable multiplication of singly charged relativistic particles (electrons) has occurred in the core. Two electron pairs of energies $\sim$ 10 BeV and $\sim$ 50 BeV (as estimated from the magnitude of the very small angles between the two tracks of the pair)[1,2] were created in the emulsions of both the first and the second plate. One of the pairs within the core gives rise to another pair; this fact supports the assumption that the additional charged particles appearing in the narrow core are fast electrons produced by high energy γ-rays (or possibly other neutral particles).

[1] M. Stearns, *Phys. Rev.* **76,** 836 (1949).
[2] H. S. Snyder and W. T. Scott, *Phys. Rev.* **76,** 220 (1949).

At least eight and probably about ten pairs are created in the core over an average path length of 0·26 radiation units of glass and emulsion; the number of γ-rays in the core must therefore at least be equal to $N_0^{(\gamma)} \sim 35$.

The energy $E = 4\gamma Mc^2$ of the primary α-particle can be estimated from the angular width of the narrow core:

$$\Delta\vartheta \sim \frac{1}{\bar{\gamma}} = (1-\bar{\beta}^2)^{\frac{1}{2}} = \left(\frac{2}{\gamma+1}\right)^{\frac{1}{2}} \sim \frac{1}{30}, \quad \gamma = 2000,$$

$$E_\alpha = 0{\cdot}8\times 10^{13}\,\text{eV}.$$

($\bar{\beta}$= velocity of the centre of mass system of a nucleon of the incident α-particle and a nucleon of the target nucleus.)[b]

A more detailed study shows that the value of γ obtained from the angular distribution of mesons in the core C does not vary too much if different not too extreme assumptions concerning the energy distribution of the mesons emitted isotropically in the centre-of-mass system are made*. The average separation of tracks in the second plate indicates that the average γ-ray energy cannot be greatly in excess of 10 BeV. It is most probable that three or four nucleons of the incident α-particle contribute to the narrow shower. Hence, if the γ-rays result from the decay of neutral mesons, whose numbers do not exceed the number of charged mesons, the collision leading to the narrow shower should not be considered to be completely inelastic, since a completely inelastic collision (with $\gamma = 2000$) would give an average γ-energy of $\sim$ 50 BeV.

Unless we assume an extremely anisotropic angular distribution of mesons in the c.m. systems (emission within a double cone of opening $\vartheta \sim 1/\bar{\gamma}$), the diffuse shower of $\sim$ 30 relativistic particles cannot result directly from the primary encounter producing the narrow core. We thus conclude that the diffuse shower is the result of secondary and tertiary meson production in the same nucleus by some of the nucleons emerging from the primary encounter.[3] To summarize we conclude

* The above energy estimate is reduced by a factor 7 if an angular distribution $f(\vartheta) \approx \cos^4\vartheta$ is assumed in the c.m. system.

[3] L. Leprince-Ringuet *et al.*, *Comptes Rendus*, **229**, 163 (1949).

that the event described (which may be the initial stage of development of an Auger shower) gives direct evidence for the multiple production of high energy γ-rays. If the γ-rays result from the decay of a neutral meson of rest mass $\sim$ 150 MeV (see following letter[c]) we can, by assuming that the γ-rays are converted immediately into electron pairs, obtain an upper limit for the proper life-time of the neutral meson of $\tau_0 \lesssim 10^{-13}$ sec.

A more detailed report will appear in *Helvetica Physica Acta*. This work was assisted by the Joint Program of the ONR and the AEC.

We are greatly indebted to Dr. R. E. Marshak for many valuable discussions.

Notes

[a] I.e. grain density, hence ionization, above minimum. Their range then indicates that they are heavy.

[b] It is supposed that when two nucleons collide there is a backward–forward symmetry of the emitted particles in the c.m. system. Transforming to the lab. system of coordinates (Lorentz transformation with relative velocity βc) results in a double-cone appearance in general. If a fraction f of the particles appear within an angle $\vartheta(f)$ of the axis, it can be shown from the symmetry assumption, if particles are relativistic in the c.m.s., that $\vartheta(f).\vartheta(1-f) = \bar{\gamma}^{-2}$, where $\bar{\gamma} = (1-\bar{\beta}^2)^{-1/2}$; in particular, $\vartheta(\frac{1}{2}) = 1/\bar{\gamma}$. $\bar{\gamma}$ is simply related to the γ of the incident nucleon.

[c] "Remarks on multiple meson and gamma-ray production" by R. E. Marshak, who argues that this evidence favours a neutral version of the π-meson, having spin zero, and pseudoscalar.

11. A Mechanism of Acquirement of Cosmic-ray Energies by Electrons[†]

W. F. G. Swann[‡]

It is shown that the growth of stellar magnetic fields such as occur in sunspots can give rise according to Faraday's law of electromagnetic induction to electric fields capable of giving electrons energies corresponding to 10^{10} volts. Moreover, the rate of acquirement of energy can reasonably be greater than the rate of loss by collisions with atoms. It would be difficult to realize energies as high as 10^{10} volts from the magnetic fields of spots such as occur on the sun. Energies corresponding to 10^9 volts are, however, within the range of possibility, and it is suggested that electrons projected from such spots may play a part in auroral phenomena. For cosmic rays one must, however, probably look to the stars for the necessary conditions. Because of the repulsion of the current responsible for the magnetic field, the electron is hurled away from the spot as it acquires its energy.

It is usually considered that the assumption of the existence of electronic energies of the order 10^{10} volts is attended with considerable theoretical difficulties. The present note is an elaboration of one of the suggestions recently made by the writer.[1] It concerns the possibility of the production of cosmic-ray energies in stellar spots, analogous to sunspots on our sun. It is known that the sunspots possess considerable magnetic fields which grow in comparatively short intervals of time. A consideration of the existing magnitudes will show that line integrals of the electric field of the order 10^{10} volts may readily arise through the Faraday law as a result of the growth of such magnetic fields. However,

[†] *Phys. Rev.* **43**, 217–220 (1933). (Received December 27th, 1932.)

[‡] Bartol Research Foundation of the Franklin Institute.

[1] In a paper under the above title presented at the Chicago Meeting of the American Physical Society, November, 1932.

the detailed working out of the matter involves certain considerations which require careful attention.

We shall consider the problem in cylindrical polar coordinates r, θ, z, in which the stellar spot is in the plane of r, θ, and the magnetic field H is symmetrical with θ and, of course, more or less perpendicular to the plane of the spot.

The problem must be treated from the standpoint of relativistic electrodynamics. It may be conveniently treated by the equations of Lagrange, or even more conveniently by the Hamiltonian equations.

The Lagrangian function for an electron in a field specified by a vector potential U and a scalar potential φ is, as is well known,

$$L = -mc^2(1-\beta^2)^{\frac{1}{2}}+(e/c)(U.u)-e\varphi,$$

where m is the rest mass, u is the velocity, and $\beta^2 = u^2/c^2$.

φ is zero in our problem, so that transforming to polar coordinates, we have

$$L = -mc^2\left(1-\frac{r^2\dot{\theta}^2}{c^2}-\frac{\dot{r}^2}{c^2}-\frac{\dot{z}^2}{c^2}\right)^{\frac{1}{2}}+\frac{e}{c}(\dot{r}U_r+r\dot{\theta}U_\theta+\dot{z}U_z).$$

The momenta P_r, P_θ, P_z, are given by

$$P_r = m\dot{r}(1-\beta^2)^{-\frac{1}{2}}+eU_r/c, \tag{1}$$

$$P_\theta = mr^2\dot{\theta}(1-\beta^2)^{-\frac{1}{2}}+erU_\theta/c, \tag{2}$$

$$P_z = m\dot{z}(1-\beta^2)^{-\frac{1}{2}}+eU_z/c. \tag{3}$$

The Hamiltonian function $\mathcal{H}$ is given by

$$\mathcal{H} = \dot{r}P_r+\dot{\theta}P_\theta+\dot{z}P_z-L \tag{4}$$

and is easily shown to be $\mathcal{H} = mc^2(1-\beta^2)^{-\frac{1}{2}}$. Solving (1), (2) and (3) for $\dot{r}$, $\dot{\theta}$, $\dot{z}$, and substituting in (4) we readily find

$$\mathcal{H} = mc^2\left\{1+\frac{1}{m^2c^2}\left[\left(P_r-\frac{e}{c}U_r\right)^2 + \left(\frac{P_\theta}{r}-\frac{e}{c}U_\theta\right)^2+\left(P_z-\frac{e}{c}U_z\right)^2\right]\right\}^{\frac{1}{2}}. \tag{5}$$

Now in our problem $U_r = U_z = 0$, and U_θ is independent of θ for a magnetic field symmetrical with regard to θ. In fact, the electron currents responsible for U are all of a circular type symmetrical about the z-axis. Thus P_θ is constant, i.e. from (2)

$$mr^2\dot{\theta}(1-\beta^2)^{-\frac{1}{2}}+erU_\theta/c = er_0U_0/c, \tag{6}$$

where r_0 and U_0 refer to the values of r and U_θ at the initial instant when $\dot{\theta} = 0$. Thus

$$\mathscr{H} = mc^2\left\{1+\frac{1}{m^2c^2}\left[P_r^2+P_z^2+\frac{e}{c}\left(\frac{r_0U_0}{r}-U_\theta\right)^2\right]\right\}^{\frac{1}{2}}. \tag{7}$$

Now $\mathscr{H}$ is explicitly a function of t through U_θ. Hence, as follows directly from the Hamiltonian equations

$$d\mathscr{H}/dt = \partial\mathscr{H}/\partial t = -(e^2/\mathscr{H})(r_0U_0/r-U_\theta)(\partial U_\theta/\partial t),$$
$$\tfrac{1}{2}d\mathscr{H}^2/dt = e^2(U_\theta-r_0U_0/r)(\partial U_\theta/\partial t). \tag{8}$$

Since $\mathscr{H} = mc^2(1-\beta^2)^{-\frac{1}{2}} = T+mc^2$, where T is the kinetic energy,

$$\tfrac{1}{2}(d/dt)(T+mc^2)^2 = e^2(U_\theta-r_0U_0/r)(\partial U_\theta/\partial t). \tag{9}$$

Hence T increases with t so long as $rU_\theta \geqslant r_0U_0$. This is certainly satisfied at the initial instant in view of the definitions of r_0 and U_0. It will be satisfied for all subsequent instants if $d(rU_\theta)/dt \geqslant 0$; i.e., if $\partial U_\theta/\partial t+(\dot{r}/r)(\partial/\partial r)(rU_\theta) \geqslant 0$; i.e., if

$$-(E_\theta-\dot{r}H_z/c) \geqslant 0. \tag{10}$$

Bearing in mind that E_θ is in the negative direction of θ for an increase of U_θ in the positive direction of θ, we see that condition (10) is simply the condition that the force due to the electric field accelerating the electron in the negative direction of θ must be always greater than the force due to the magnetic field which acts in the opposite sense when $\dot{r}$ is positive. If (10) is satisfied always, the electron will never loop around in its path since to so loop $-(E_\theta-\dot{r}H_z/c)$ would have to be less than zero in order to balance the centrifugal acceleration at the instant when $\dot{\theta}$ was zero. Since $\dot{r}$ can never exceed c, the condition for continual increase of energy will always be satisfied if $|E| > |H_z|$. If there is anything like a uniform growth of magnetic flux with time, E will certainly be greater than H initially, and we shall find that there is time for high electronic energy to be acquired before H has grown to a magnitude sufficiently large to prohibit further increase.

If the electron starts from rest, at $t = 0$ and if the magnetic field is zero then we have $U_0 = 0$, and

$$\tfrac{1}{2}(d/dt)(T+mc^2)^2 = e^2U_\theta(\partial U_\theta/\partial t). \tag{11}$$

So far we have made no assumption as to the way in which the magnetic field H depends upon r. Since $H = \text{curl } U$, Stokes theorem applied to an annulus $2\pi r\, dr$ perpendicular to z gives

$$2\pi r H_z = \partial(2\pi r U_\theta)/\partial r. \tag{12}$$

If, for purposes of illustration we take a case where H_z is inversely proportional to r, we shall enjoy the advantage of having U_θ independent of r as well as of θ. We shall consequently write

$$H_z = H_0 Rt/r\tau \tag{13}$$

where τ is the time taken for it to acquire a field intensity H_0 at some assigned radius R comparable with the radius of the spot.

With U_θ independent of r and $U_r = 0$, $\mathscr{H}$ is independent of r, so that P_r is constant with time, which, through (1) shows that $\dot{r}$ is zero for all time since it is zero initially. Thus the electron describes a circular path. As a matter of fact an examination of the equations shows, as might be expected, that, under these conditions, the centrifugal acceleration of the electron and the magnetic deflecting force increase together with the time in such a manner as to keep each other balanced.

Thus, with the assumption involved in (13), (12) gives $U_\theta = H_0Rt/\tau$ and (11) gives

$$(d/dt)(T+mc^2)^2 = 2e^2H_0^2R^2t/\tau^2 \tag{14}$$

and $(T+mc^2)^2 = e^2H_0^2R^2t^2/\tau^2+m^2c^4$. Since T is zero when t is zero.

For the energies in which we are interested, energies of the order 10^{10} volts, mc^2 is negligible, so that $T = eH_0Rt/\tau$ or, expressed in volts,

$V = 300 H_0 R t/\tau$. If $R = 3 \times 10^{10}$, $H_0 = 2000$, $\tau = 10^6$, $t = 1$ second, we find $V = 2 \times 10^{10}$ volts.

The conclusion is then to the effect that in one second during the initial formation of a stellar spot about 50 times the earth's diameter and of such character as to give rise to a magnetic field of 2000 gauss in 12 days, an electron can acquire a velocity comparable with 2×10^{10} volts. The electric field E is given by

$$\int_0^R \frac{H_0 R}{r\tau} 2\pi r \, dr = 2\pi R c E,$$

so that $E = H_0 R/c\tau$ and the magnetic field at the end of a time t is $H_0 t/\tau$. Hence in the present instance, the magnetic field would only just have attained a value equal to the electric field by the time the electron had received 2×10^{10} volts velocity. In the present case, the fact that $\dot{r}$ is zero insures that the electron would go on acquiring energy continually, even after the magnetic field had become equal to or greater than the electric field; but this feature is special to our particular assumption as to the dependence of the magnetic field upon r. Naturally our particular assumption regarding H is mainly relevant only to the mathematical simplification of the problem and not to its essential features. As a matter of fact the magnetic field we have assumed would not have zero curl, and would require the existence of circular electron currents in the regions where H has the property specified. Difficulties regarding the infinite value of H at $r = 0$ are irrelevant since we only require the variation of H with r postulated to exist over a very thin range of r. If we had a uniform magnetic field instead of one of the type assumed, the electron would not describe a circular orbit but would diminish its distance from the origin in its path. This complicates the details of our calculation but not the order of magnitude of the results.

We have given no attention to the motion of the electron parallel to z. It is obvious that since the electronic motion takes place in the opposite sense to that of the increasing electronic currents which are producing the magnetic field, the electron will be repelled from them, and so will shoot out into space, which is a desirable consummation from the point of view of our theory.

An examination of the foregoing calculation shows that it would be difficult to secure energies of 10^{10} volts from spots of a size such as occur on our sun. However, energies one-tenth of this amount are within the range of possibility. This fact is particularly interesting in suggesting that while we may have to look to stars other than our sun for sources of cosmic rays, the spots on the sun may be able to furnish electrons such as could play a part in auroral phenomena, since the position of the auroral zone suggests energies of the order of 10^9 volts for the electrons producing auroras. It is moreover a significant fact in this connection that correlations between sunspot activity and auroral and terrestrial magnetic phenomena have been observed. Another matter of significance arises from the fact that although, for purposes of visualization of the situation, we have imagined a spot which grows 2000 gauss in 10^6 seconds, as a matter of fact the electronic energy of 2×10^{10} is acquired before the spot has been growing for more than one second, or before the field has attained a value of more than 2×10^{-3} gauss in our example. Naturally it is not intended to imply that a situation exists in nature in which the problem we have cited is duplicated in all its details. All that it is intended to show is that with magnetic fields extending over large areas in stellar spots, relatively small rates of changes of these fields existing for brief periods can give rise to large electronic energies. Our problem has been designed to illustrate the feasibility of the idea in one particular case.

Finally, we must consider the possibility of loss of energy by collision. The work done on the electron over its mean free path λ is $E\lambda e$. In our present example $E = RH_0/\tau c = 2\times10^{-3}$ e.s.u. Hence the work estimated in volts is $10^{-3}\times600\lambda = 0{\cdot}6\lambda$ volt. Hence if λ were greater than 50 cm the electron would acquire more energy over its mean free path than it would lose at collision, and the energy would continue to increase with time. The pressure above the surface of a star falls off very rapidly with distance and is probably far below that corresponding to a mean free path of 50 cm in regions which are sufficienty near to the surface to experience the magnetic fields of the spots. It is probable, therefore, that the efficiency of the process of accumulation of energy is but little reduced by collisions.

12. On the Origin of the Cosmic Radiation[†]

ENRICO FERMI[‡]

A theory of the origin of cosmic radiation is proposed according to which cosmic rays are originated and accelerated primarily in the interstellar space of the galaxy by collisions against moving magnetic fields. One of the features of the theory is that it yields naturally an inverse power law for the spectral distribution of the cosmic rays. The chief difficulty is that it fails to explain in a straightforward way the heavy nuclei observed in the primary radiation.

I. Introduction

In recent discussions on the origin of the cosmic radiation E. Teller[1] has advocated the view that cosmic rays are of solar origin and are kept relatively near the sun by the action of magnetic fields. These views are amplified by Alfvén, Richtmyer, and Teller.[2] The argument against the conventional view that cosmic radiation may extend at least to all the galactic space is the very large amount of energy that should be present in form of cosmic radiation if it were to extend to such a huge space. Indeed, if this were the case, the mechanism of acceleration of the cosmic radiation should be extremely efficient.

I propose in the present note to discuss a hypothesis on the origin of cosmic rays which attempts to meet in part this objection, and according to which cosmic rays originate and are accelerated primarily in the interstellar space, although they are assumed to be prevented by magnetic fields from leaving the boundaries of the galaxy. The main process

† *Phys. Rev.* **75**, 1169–1174 (1949). (Received January 3rd, 1949.)
‡ Institute for Nuclear Studies, University of Chicago.
[1] Nuclear Physics Conference, Birmingham, 1948.
[2] Alfvén, Richtmyer, and Teller, *Phys. Rev.*, to be published.

of acceleration is due to the interaction of cosmic particles with wandering magnetic fields which, according to Alfvén, occupy the interstellar spaces.

Such fields have a remarkably great stability because of their large dimensions (of the order of magnitude of light years), and of the relatively high electrical conductivity of the interstellar space. Indeed, the conductivity is so high that one might describe the magnetic lines of force as attached to the matter and partaking in its streaming motions. On the other hand, the magnetic field itself reacts on the hydrodynamics[3] of the interstellar matter giving it properties which, according to Alfvén, can pictorially be described by saying that to each line of force one should attach a material density due to the mass of the matter to which the line of force is linked. Developing this point of view, Alfvén is able to calculate a simple formula for the velocity V of propagation of magneto-elastic waves:

$$V = H/(4\pi\varrho)^{\frac{1}{2}}, \tag{1}$$

where H is the intensity of the magnetic field and ϱ is the density of the interstellar matter.

One finds according to the present theory that a particle that is projected into the interstellar medium with energy above a certain injection threshold gains energy by collisions against the moving irregularities of the interstellar magnetic field. The rate of gain is very slow but appears capable of building up the energy to the maximum values observed. Indeed one finds quite naturally an inverse power law for the energy spectrum of the protons. The experimentally observed exponent of this law appears to be well within the range of the possibilities.

The present theory is incomplete because no satisfactory injection mechanism is proposed except for protons which apparently can be regenerated at least in part in the collision processes of the cosmic radiation itself with the diffuse interstellar matter. The most serious difficulty is in the injection process for the heavy nuclear component of the radiation. For these particles the injection energy is very high and the injection mechanism must be correspondingly efficient.

[3] H. Alfvén, *Arkiv Mat. f. Astr., o. Fys.* **29B**, 2 (1943).

II. The motions of the interstellar medium

It is currently assumed that the interstellar space of the galaxy is occupied by matter at extremely low density, corresponding to about one atom of hydrogen per cc, or to a density of about 10^{-24} g/cc. The evidence indicates, however, that this matter is not uniformly spread, but that there are condensations where the density may be as much as ten or a hundred times as large and which extend to average dimensions of the order of 10 parsec. (1 parsec = $3 \cdot 1 \times 10^{18}$ cm = 3·3 light years.) From the measurements of Adams[4] on the Doppler effect of the interstellar absorption lines one knows the radial velocity with respect to the sun of a sample of such clouds located at not too great distance from us. The root mean square of the radial velocity, corrected for the proper motion of the sun with respect to the neighboring stars, is about 15 km/sec. We may assume that the root-mean-square velocity is obtained by multiplying this figure by the square root of 3, and is therefore about 26 km/sec. Such relatively dense clouds occupy approximately 5% of the interstellar space.[5]

Much less is known of the much more dilute matter between such clouds. For the sake of definiteness in what follows, the assumption will be made that this matter has a density of the order of 10^{-25}, or about 0·1 hydrogen atoms per cc. Even fairly extensive variations on this figure would not very drastically alter the qualitative conclusions. If the assumption is made that most of this material consists of hydrogen atoms, it is to be expected that most of the hydrogen will be ionized by the photoelectric effect of the stellar light. Indeed, one can estimate that some kind of dissociation equilibrium is established under average interstellar conditions, outside the relatively dense clouds, for which

$$n_+^2/n_0 \approx (T_1)^{\frac{1}{2}}, \tag{2}$$

where n_+ and n_0 are the concentrations of ions and neutral atoms per cc, and T_1 is the absolute kinetic temperature in degrees K. Putting in this formula $n_+ = 0 \cdot 1$, one finds that the fraction of undissociated

[4] W. S. Adams, *A.p.J.* **97,** 105 (1943).
[5] B. Stromgren, *A.p.J.* **108,** 242 (1948).

atoms is of the order of 1%, even assuming a rather low kinetic temperature of the order of 100°K.[a]

It is reasonable to assume that this very low density medium will have considerable streaming motions, since it will be kept stirred by the moving heavier clouds passing through it. In what follows, a root-mean-square velocity of the order of 30 km/sec will be assumed. According to Alfvén's picture, we must assume that the kinetic energy of these streams will be partially converted into magnetic energy, that indeed, the magnetic field will build up to such a strength that the velocity of propagation of the magneto-elastic waves becomes of the same order of magnitude as the velocity of the streaming motions. From (1) it follows then that the magnetic field in the dilute matter is of the order of magnitude of 5×10^{-6} gauss, while its intensity is probably greater in the heavier clouds. The lines of force of this field will form a very crooked pattern, since they will be dragged in all directions by the streaming motions of the matter to which they are attached. They will, on the other hand, tend to oppose motions where two portions of the interstellar matter try to flow into each other, because this would lead to a strengthening of the magnetic field and a considerable increase of magnetic energy. Indeed, this magnetic effect will have the result to minimize what otherwise would be extremely large friction losses which would damp the streaming motions and reduce them to disordered thermal motions in a relatively short time.

III. Acceleration of the cosmic rays

We now consider a fast particle moving among such wandering magnetic fields. If the particle is a proton having a few BeV energy, it will spiral around the lines of force with a radius of the order of 10^{12} cm until it "collides" against an irregularity in the cosmic field and so is reflected, undergoing some kind of irregular motion. On a collision both a gain or a loss of energy may take place. Gain of energy, however, will be more probable than loss. This can be understood most easily by observing that ultimately statistical equilibrium should be

[a] Superscript letters refer to Notes at end of articles.

established between the degrees of freedom of the wandering fields and the degrees of freedom of the particle. Equipartition evidently corresponds to an unbelievably high energy. The essential limitation, therefore, is not the ceiling of energy that can be attained, but rather the rate at which energy is acquired. A detailed discussion of this process of acceleration will be given in section VI. An elementary estimate can be obtained by picturing the "collisions" of the particles against the magnetic irregularities as if they were collisions against reflecting obstacles of very large mass, moving with disordered velocities averaging to $V = 30$ km/sec. Assuming this picture, one finds easily that the average gain in energy per collision is given as order of magnitude by

$$\delta w = B^2 w, \tag{3}$$

where w represents the energy of the particle inclusive of rest energy, and $B = V/c \approx 10^{-4}$. This corresponds, therefore, for a proton to an average gain of 10 volts per collision in the non-relativistic region, and higher as the energy increases. It follows that except for losses the energy will increase by a factor e every 10^8 collisions. In particular, a particle starting with non-relativistic energy will attain, after N collisions, an energy

$$w = Mc^2 \exp(B^2 N). \tag{4}$$

Naturally, the energy can increase only if the losses are less than the gain in energy. An estimate to be given later (see section VII) indicates that the ionization loss becomes smaller than the energy gain for protons having energy of about 200 MeV. For higher energy the ionization loss practically becomes negligible. We shall discuss later the injection mechanism.

IV. Spectrum of the cosmic radiation

During the process of acceleration a proton may lose most of its energy by a nuclear collision. This process is observed as absorption of primary cosmic radiation in the high atmosphere and occurs with a mean free path of the order of magnitude of 70 g/cm², corresponding to a cross-section of about

$$\sigma_{abs} \approx 2{\cdot}5\times 10^{-26}\ \text{cm}^2 \tag{5}$$

per nucleon.

In a collision of this type most of the kinetic energy of the colliding nucleons is probably converted into energy of a spray of several mesons.

It is reasonable to assume that the cosmic rays will occupy with approximately equal density all the interstellar space of the galaxy. They will be exposed, therefore, to the collisions with matter of an average density of 10^{-24}, leading to an absorption mean free path

$$\Lambda = 7\times 10^{25}\ \text{cm}. \tag{6}$$

A particle traveling with the velocity of light will traverse this distance in a time

$$T = \Lambda/c = 2\times 10^{15}\ \text{sec}. \tag{7}$$

or about 60 million years.

The cosmic-ray particles now present will therefore, in the average, have this age. Some of them will have accidentally escaped destruction and be considerably older. Indeed, the absorption process can be considered to proceed according to an exponential law. If we assume that original particles at all times have been supplied at the same rate, we expect the age distribution now to be

$$\exp(-t/T)\,dt/T. \tag{8}$$

During its age t, the particle has been gaining energy. If we call τ the time between scattering collisions, the energy acquired by a particle of age t will be

$$w(t) = Mc^2 \exp(B^2 t/\tau). \tag{9}$$

Combining this relationship between age and energy with the probability distribution of age given previously, one finds the probability distribution of the energy. An elementary calculation shows that the probability for a particle to have energy between w and $w+dw$ is given by

$$\pi(w)\,dw = (\tau/B^2T)(Mc^2)^{\tau/B^2T}\,dw/w^{1+\tau/B^2T}. \tag{10}$$

It is gratifying to find that the theory leads naturally to the conclusion that the spectrum of the cosmic radiation obeys an inverse power law. By comparison of the exponent of this law with the one known from cosmic-ray observations, that is, about 2·9, one finds a relationship which permits one to determine the interval of time τ between collisions. Precisely, one finds: $2{\cdot}9 = 1+\tau/B^2T$, from which follows

$$\tau = 1{\cdot}9B^2T. \tag{11}$$

Using the previous values of B and T, one finds $\tau = 4\times 10^7 \approx 1{\cdot}3$ years. Since the particles travel with approximately the velocity of light, this corresponds to a mean distance between collisions of the order of a light year, or about 10^{18} cm. Such a collision mean free path seems to be quite reasonable.

The theory explains quite naturally why no electrons are found in the primary cosmic radiation. This is due to the fact that at all energies the rate of loss of energy by an electron exceeds the gain. At low energies, up to about 300 MeV, the loss is mainly due to ionization. Above this energy radiative losses due to the acceleration of the electrons in the interstellar magnetic field play the dominant role. This last energy loss is instead quite negligible for protons. Also, the inverse Compton effect discussed by Feenberg and Primakoff[6] will contribute to eliminate high energy electrons.

V. The injection mechanism. Difficulties with the injection of heavy nuclei

In order to complete the present theory, the injection mechanism should be discussed.

In order to keep the cosmic radiation at the present level it is necessary to inject a number of protons of at least 200 MeV, to compensate for those that are lost by the absorption process. According to recent evidence,[7] the primary cosmic radiation contains not only protons but

[6] E. Feenberg and H. Primakoff, *Phys. Rev.* **73**, 449 (1948).

[7] Freier, Lofgren, Ney, and Oppenheimer, *Phys. Rev.* **74**, 1818 (1948); H. L. Bradt and B. Peters, *Phys. Rev.* **74**, 1828 (1948).

also some relatively heavy nuclei. Their injection energy is much higher than that of protons, primarily on account of their large ionization loss. (See further section VII.) Such high energy protons and heavier nuclei conceivably could be produced in the vicinity of some magnetically very active star.[8] To state this, however, merely means to shift the difficulty from the problem of accelerating the particles to that of injecting them unless a more precise estimate can be given for the efficiency of this or of some equivalent mechanism. With respect to the injection of heavy nuclei I do not know a plausible answer to this point.

For the production of protons, however, one might consider also a simple mechanism which, if the present theory is at all correct in its general features, should be responsible for at least a large fraction of the total number of protons injected. According to this mechanism the cosmic radiation regenerates itself as follows. When a fast cosmic-ray proton collides in the interstellar space against a proton nearly at rest, a good share of the energy will be lost in the form of a spray of mesons, and two nucleons will be left over with energy much less than that of the original cosmic ray. Estimates indicate that in some cases both particles may have an energy left over above the injection threshold of 200 MeV, in some cases one and in some cases none. We can introduce a reproduction factor k, defined as the average number of new protons above the injection energy arising in a collision of an original cosmic-ray particle. As in a chain reaction, if k is greater than one the over-all number of cosmic rays will increase; if k is less than one it will decrease; if k is equal to one it will stay level.

Apparently the reproduction factor under interstellar conditions is rather close to one. This is perhaps not a chance, but may be due in part to the following self-stabilizing mechanism. The motions of the interstellar matter are not quite conservative, in spite of the reduced friction, caused by the magnetic fields. One should assume, therefore, that some source is present which steadily delivers kinetic energy into the streaming motions of the interstellar matter. Probably such a source of energy ultimately involves conversion of energy from the

[8] See for example W. F. G. Swann, *Phys. Rev.* **43**, 217 (1933) and Horace W. Babcock, *Phys. Rev.* **74**, 489 (1948).

large supplies in the interior of the stars. The motions of the interstellar medium are in a dynamic equilibrium between the energy delivered by this source and the energy losses caused by friction and other causes. In this balance the amount of energy transferred by the interstellar medium to cosmic radiation is by no means irrelevant, since the total cosmic ray energy is comparable to the kinetic energy of the streaming, irregular motions of the galaxy. One should expect, therefore, that if the general level of the cosmic radiation should increase, the kinetic energy of the interstellar motion would decrease, and vice versa. The reproduction factor depends upon the density. As the density increases, the ionization losses will increase proportionally to it. This tends to increase the injection energy and consequently to decrease the reproduction factor. On the other hand, also, the rate of energy gained will change by an amount which is hard to define unambiguously. One might perhaps assume, however, that the velocity of the wandering magnetic fields increases with the $\frac{1}{3}$ power of the density, as would correspond to the virial theorem, and that the collision mean free path is inversely proportional to the $\frac{1}{3}$ power of the density, as one might get from geometrical similitude. One would find that the rate of energy increase is proportional only to the $\frac{2}{3}$ power of the density. The net effect is an increase of the injection energy and a decrease of the reproduction factor with increasing density. If the reproduction factor had been initially somewhat larger than one, the general level of the cosmic radiation would increase, draining energy out of the kinetic energy of the galaxy. This would determine a gravitational contraction which would increase the density and decrease k until the stable value of one is reached. The opposite would take place if k initially had been considerably less than one.

But even if this stabilizing mechanism is not adequate to keep the reproduction factor at the value one, and therefore an appreciable change in the general level of the cosmic radiation occurs over periods of hundreds of millions of years, the general conclusions reached in section IV would not be qualitatively changed. Indeed, if k were somewhat different from one, the general level of the cosmic radiation would increase or decrease exponentially, depending on whether k is larger than or less than one. Consequently the number of cosmic particles injected according to the mechanism that has been dicussed will not be constant in time but will vary exponentially. Combining this exponential variation with the exponential absorption (8), one still finds an exponential law for the age distribution of the cosmic particles at the present time, the only difference being that the period of this exponential will be changed by a small numerical factor.

The injection mechanism here proposed appears to be quite straightforward for protons, but utterly inadequate to explain the abundance of the heavy nuclei in the primary cosmic radiation. The injection energy of these particles is of several BeV, and it is difficult to imagine a secondary effect of the cosmic radiation on the diffuse interstellar matter which might produce this type of secondary with any appreciable probability. One might perhaps assume that the heavy particles originate at the fringes of the galaxy where the density is probably lower and the injection energy is therefore probably smaller. This, however, would require extreme conditions of density which are not easily justifiable. It seems more probable that heavy particles are injected by a totally different mechanism, perhaps as a consequence of the stellar magnetism.[8]

If such a mechanism exists one would naturally expect that it would inject protons together with heavier nuclei. The protons and perhaps to a somewhat lesser extent the α-particles would be further increased in numbers by the "chain reaction" which in this case should have $k < 1$. Indeed their number would be equal to the number injected during the lifetime T increased by the factor $1/(1-k)$. Heavy particles instead would slowly gain or slowly lose energy according to whether their initial energy is above or below the injection threshold. They would, however, have a shorter lifetime than protons because of the presumably larger destruction cross-section. Their number should be approximately equal to the number injected during their lifetime.

One should remark in this connection that a consequence of the present theory is that the energy spectrum of the heavy nuclei of the cosmic radiation should be quite different from the spectrum of the protons, since the absorption cross-section for a heavy particle is presumably several times larger than that of a proton. One would expect, therefore, that the average age of a heavy particle is shorter than the age of a proton, which leads to an energy spectrum decreasing much more rapidly with energy for a heavy particle than it does for protons. An experimental check on this point should be possible.

VI. Further discussion of the magnetic acceleration

In this section the process of acceleration of the cosmic-ray protons by collision against irregularities of the magnetic field will be discussed in somewhat more detail than has been done in section III.

The path of a fast proton in an irregular magnetic field of the type that we have assumed will be represented very closely by a spiraling motion around a line of force. Since the radius of this spiral may be of the order of 10^{12} cm, and the irregularities in the field have dimensions of the order of 10^{18} cm, the cosmic ray will perform many turns on its spiraling path before encountering an appreciably different field intensity. One finds by an elementary discussion that as the particle approaches a region where the field intensity increases, the pitch of the spiral will decrease. One finds precisely that

$$\sin^2 \vartheta / H \approx \text{constant}, \tag{12}$$

where ϑ is the angle between the direction of the line of force and the direction of the velocity of the particle, and H is the local field intensity. As the particle approaches a region where the field intensity is larger, one will expect, therefore, that the angle ϑ increases until $\sin \vartheta$ attains the maximum possible value of one. At this point the particle is reflected back along the same line of force and spirals backwards until the next region of high field intensity is encountered. This process will be called a "Type A" reflection. If the magnetic field were static, such a reflection would not produce any change in the kinetic energy of the particle. This is not so, however, if the magnetic field is slowly variable. It may happen that a region of high field intensity moves toward the cosmic-ray particle which collides against it. In this case, the particle will gain energy in the collision. Conversely, it may happen that the region of high field intensity moves away from the particle. Since the particle is much faster, it will overtake the irregularity of the field and be reflected backwards, in this case with loss of energy. The net result will be an average gain, primarily for the reason that head-on collisions are more frequent than overtaking collisions because the relative velocity is larger in the former case.

Somewhat similar processes take place when the cosmic-ray particle spirals around a curve of the line of force as outlined in Figure 1 ("Type *B*" reflection). Here again, the energy of the particle would

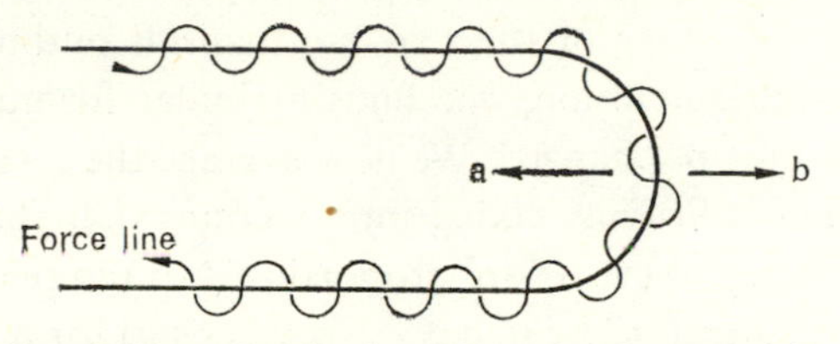

FIG. 1. Type *B* reflection of a cosmic-ray particle.

not change if the magnetic field were static. On the other hand, the lines of force partake of the streaming motions of the matter, and it may happen that the line of force at the bottom of the curve moves in the direction indicated by the arrow *a*, or that it moves in the direction indicated by the arrow *b*. In the former case there will be an energy gain (head-on collision) while in case *b* (overtaking collision) there will be an energy loss. Gain and loss, however, do not average out completely, because also in this case a head-on collision is slightly more probable than an overtaking collision due to the greater relative velocity.

The amount of energy gained or lost in a collision of the two types described can be estimated with a simple argument of special relativity, without any reference to the detailed mechanism of the collision. In the frame of reference in which the perturbation of the field against which the collision takes place is at rest, there is no change of energy of the particle. The change of energy in the rest frame of reference is obtained, therefore, by first transforming initial energy and momentum from the rest frame to the frame of the moving perturbation. In this frame an elastic collision takes place whereby the momentum changes direction and the energy remains unchanged. Transforming back to the frame of reference at rest, one obtains the final values of energy and momentum. This procedure applied to a head-on collision, gives the following result,

$$\frac{w'}{w} = \frac{1+2B\beta\cos\vartheta+B^2}{1-B^2}, \tag{13}$$

where βc is the velocity of the particle, ϑ is the angle of inclination of the spiral, and Bc is the velocity of the perturbation. It is assumed that the collision is such as to produce a complete reversal of the spiraling direction by either of the two mechanisms outlined previously. For an overtaking collision, one finds a similar formula except that the sign of B must be changed. We now average the results of head-on and overtaking collisions, taking into account that the probabilities of these two types of events are proportional to the relative velocities and are given therefore by $(\beta\cos\vartheta+B/2\beta\cos\vartheta)$ for a head-on collision and $(\beta\cos\vartheta-B/2\beta\cos\vartheta)$ for an overtaking collision. The result

TABLE 1. ENERGY LOSS PER g/cm² OF MATERIAL TRAVERSED

Energy	Loss/g/cm²	Gain/g/cm²
10^7 eV	94×10^6 eV	$7{\cdot}8\times10^6$ eV
10^8	15×10^6	$8{\cdot}6\times10^6$
10^9	$4{\cdot}6\times10^6$	$16{\cdot}1\times10^6$
10^{10}	$4{\cdot}6\times10^6$	91×10^6

for the average of $\ln(w'/w)$ up to terms of the order of B^2 is:

$$\langle\ln(w'/w)\rangle_{\text{Av}} = 4B^2-2B^2\beta^2\cos^2\vartheta \tag{14}$$

which confirms the order of magnitude for the average gain of energy adopted in section III.

As a result of the extreme complication of the magnetic field and of its motion, it does not appear practical to attempt an estimate by more than the order of magnitude.

One might expect that after a relatively short time the angle ϑ will be reduced to a fairly low value so that Type A reflections will become infrequent. This is due to the fact that when ϑ is large, fairly large increases in energy and decreases of ϑ may occur, if the particle should be caught between two regions of high field moving against each other along a force line. One can prove that Type B reflections change gradually and rather slowly the average pitch of the spiral. It appears, therefore, that except for the beginning of the acceleration processes the Type A will not give as large a contribution as one otherwise might expect.

VII. Estimate of the injection energy

Acceleration of a cosmic-ray particle will not be possible unless the energy gain is greater than the ionization loss. Since this last is very large for protons of low velocity, only protons above a certain energy threshold will be accelerated. In section III a value of 200 MeV for this "injection energy" has been given; a justification for this assumed value will be given now. In estimating the injection energy we will assume that the particle, during its acceleration, finds itself both inside relatively dense clouds and in the more dilute material outside of the clouds for lengths of time proportional to the volumes of these two regions. The ionization loss will be due, therefore, to a material of an average density equal to the average density of the interstellar matter, which has been assumed to be 10^{-24} g/cm^2, consisting mostly of hydrogen. In Table 1 the energy loss per g/cm^2 of material traversed is given as a function of the energy of the proton. In the third column of the table the corresponding energy gain is given. It is seen that the loss exceeds the gain for particles of energy less than about 200 MeV, as has been stated.

A similar estimate yields for the acceleration of α-particles an injection energy of about 1 BeV, for the acceleration of oxygen nuclei the initial energy required is about 20 BeV, and for an iron nucleus it would amount to about 300 BeV. As already stated, it does not appear probable that the heavy nuclei found in the cosmic radiation are accelerated by the process here described, unless they should originate at some place in the galaxy where the interstellar material is extremely dilute.

I would like to acknowledge the help that I had from several discussions with E. Teller on the relative merits of the two opposing views that we are presenting. I learned many facts on cosmic magnetism from a discussion with H. Alfvén, on the occasion of his recent visit to Chicago. The views that he expressed then were quite material in influencing my own ideas on the subject.

Note

[a] The degree of ionization is now known to be very much less than this, except close to hot stars, but the conductivity seems to be enough to maintain magnetic fields of the strength envisaged.

13. Cosmic Radiation and Radio Stars†

H. ALFVÉN and N. HERLOFSON‡

THE normal radio wave emission from the sun amounts to 10^{-17} of the heat radiation, and increases during bursts[1] to as much as 10^{-13}. If a radio star, e.g. the source in Cygnus, is situated at a distance of 100 light years,[a] its radio emission is of the order of 10^{-4} of the heat radiation of our sun. It is very unlikely that the atmosphere of any star could be so different from the sun's atmosphere as to allow a radio emission which is 10^9 to 10^{13} times greater, and it seems therefore to be excluded that the source could be as small as a star. The recent discovery[2] that the intensity variations of radio stars is a "twinkling" makes it possible to assume larger dimensions.

Ryle has suggested that there should be a connection between radio stars and cosmic radiation.[3]

According to a recent development of Teller and Richtmyer's theory of cosmic radiation, the sun should be surrounded by a "trapping field" of the order 10^{-6} to 10^{-5} gauss, which confines the cosmic rays to a region with dimensions of about 10^{17} cm (0·1 light year).[4] It is likely that almost every star has a cosmic radiation of its own, trapped

† *Phys. Rev.* **78,** 616 (1950). (Received April 17th, 1950.)

‡ Royal Institute of Technology, Stockholm, Sweden.

[1] T. S. Hey, *M.N.R.A.S.* **109,** 179 (1949).

[a] Supercript letters refer to Notes at end of articles.

[2] Smith, Little and Lovell, *Nature*, **165,** 424 (1950).

[3] M. Ryle, *Proc. Phys. Soc. London,* **A62,** 491 (1949).

[4] Alfvén, Richtmyer and Teller, *Phys. Rev.* **75,** 892 (1949); R. D. Richtmyer and E. Teller, *Phys. Rev.* **75,** 1729 (1949); H. Alfvén, *Phys. Rev.* **75,** 1732 (1949) and **77,** 375 (1950).

in a region of similar size. We suggest that the radio star emission is produced by cosmic-ray electrons in the trapping field of a star.

Electrons with an energy $W \gg m_0c^2$ moving in a magnetic field H radiate at a rate

$$-dW/dt = (2e^2/3c)\omega_0^2\alpha^2, \tag{1}$$

where[5] $\omega_0 = eH/m_0c$ is the gyro-frequency corresponding to the rest mass m_0, and $\alpha = W/m_0c^2$. Most of the energy is emitted with a frequency of the order

$$\nu = \omega_0\alpha^2/2\pi = 2{\cdot}8\times10^6\, H\alpha^2 \text{ sec}^{-1}. \tag{2}$$

As soon as the energy is much higher than the rest energy the emitted frequency becomes much higher than the gyro-frequency, a phenomenon which is observed in large synchrotrons, where the electron beam emits visual light.[6]

According to (2) an emission of radio waves of 100 Mc/sec requires

$$H\alpha^2 = 36 \text{ gauss}. \tag{3}$$

The acceleration process of cosmic radiation should accelerate electrons as well as positive particles. In the solar environment the electron component is eliminated by Compton collisions with solar light quanta as discussed by Feenberg and Primakoff.[7] In the neighbourhood of a star which does not emit much light, the electrons would be accelerated until their energy is so high that they radiate. The wavelength falls in the metre band if, for example $\alpha = 300$ ($W = 1{\cdot}5\times10^8$ eV) and $H = 3\times10^{-4}$ gauss. This field is about 100 times the estimated strength of the sun's trapping field. As the strength is determined by the "interstellar wind", a radio star should be situated in an interstellar cloud moving rather rapidly relative to the star.

In order to account for the total energy emitted by a radio star we must suppose either that the radio emission is a transitory phenome-

[5] D. Iwanenko and I. Pomeranchuk, *Phys. Rev.* **65**, 343 (1944); J. Schwinger, *Phys. Rev.* **75**, 1912 (1949).

[6] Elder, Langmuir and Pollock, *Phys. Rev.* **74**, 52 (1948).

[7] E. Feenberg and H. Primakoff, *Phys. Rev.* **73**, 449 (1948).

non, lasting a time which is short compared to the lifetime of cosmic rays in the trapping field (10^8 years), or that the cosmic-ray acceleration close to the star is supplemented by a Fermi process[8] further out in the trapping field.

According to the views presented here, a radio star must not emit very much light, and should be situated in an interstellar cloud. This would explain why it is so difficult to find astronomical objects associable with the radio stars.

Note

[a] Cygnus A is actually $\sim 7\times10^6$ times further away than this, but its radio to heat ratio is unaffected. The Crab nebula is closer in scale to what was envisaged, though still somewhat larger.

[8] E. Fermi, *Phys. Rev.* **75**, 1169 (1949).

14. Abundance of Lithium, Beryllium, Boron and Other Light Nuclei in the Primary Cosmic Radiation and the Problem of Cosmic-ray Origin†

H. L. BRADT‡ and B. PETERS§

A combination of sensitive and insensitive photographic emulsions is used to improve the accuracy of measurement of the specific energy loss of relativistic nuclei. The method is applied to the determination of the charge spectrum of the primary cosmic radiation at geomagnetic latitude $\lambda = 30°$ and results in nearly complete resolution between neighboring elements up to $Z \sim 14$. It is shown that lithium, beryllium, and boron nuclei are almost or entirely absent in the primary cosmic ray beam. Since nuclear disintegrations of energetic heavy ions frequently lead to fragments of $3 \leqslant Z \leqslant 5$ it is concluded that most heavy primary nuclei do not suffer nuclear collisions between the time of acceleration and arrival and that this time must be less than $10^6/\varrho$ years, where ϱ is the density of atoms along the trajectory.

This result seems to be in disagreement with several recent theories on the origin of cosmic rays and raises some general difficulties for theories assuming galactic origin.

I. Introduction

In recent years various new theories of the origin of cosmic rays have been proposed.[1–6] These theories may be classified according to the proposed mechanism for acceleration of the nuclei of the primary

† *Phys. Rev.* **80,** 943–953 (1950). (Received July 28th, 1950.) This work was assisted by the joint program of the ONR and AEC.

‡ On May 24th, 1950, Helmut L. Bradt died unexpectedly, as this work, to which he devoted the last months of his life, was nearing completion.

§ University of Rochester, Rochester, New York.

[1] E. Fermi, *Phys. Rev.* **75,** 1169 (1949).

[2] R. D. Richtmyer and E. Teller, *Phys. Rev.* **75,** 1729 (1949).

[3] H. Alfvén, *Phys. Rev.* **75,** 1732 (1949); **77,** 375 (1950).

[4] L. Spitzer, Jr., *Phys. Rev.* **75,** 583 (1949).

[5] A. Unsöld, *Zeits. f. Astrophys,* **26,** 176 (1949).

[6] E. McMillan, *Phys. Rev.* **79,** 498 (1950)

cosmic radiation (electromagnetic acceleration, radiation pressure, etc.) or according to the assumptions made concerning the source region. The question of whether cosmic rays are of solar, galactic, or extragalactic origin may ultimately find a conclusive answer on the basis of observations on the dependence of cosmic-ray intensity on sidereal or solar time.[7–9] Another classification, which is independent of the assumed specific mechanism of acceleration and not necessarily uniquely correlated with the problem of the source region may be based on the answer to the following question: What happens to the nuclei of the cosmic radiation after they have been accelerated but before they hit the earth's atmosphere? In particular, have the cosmic-ray nuclei, on their voyage through space, traversed an amount of interstellar matter sufficient to establish equilibrium with their collision products? A theory based on the assumption of magnetic fields sufficiently strong to retain the particles in a closed region (containing the source and the earth) until they collide with interstellar matter, in gaseous or dust form, would imply a positive answer. A negative answer would, for instance, result from a theory which assumes that the majority of the primary nuclei, after being accelerated in a given region, either escape from that region, or are adiabatically decelerated or collide with bodies sufficiently large to absorb both the primaries and their break-up products and that these processes take place in a time short compared with the time between collisions with atoms in interstellar space. A negative answer would also result from a non-equilibrium theory like that proposed by Lemaître.[10] It is the purpose of this paper to show how an empirical answer to the question formulated above can be obtained and to present the results of an investigation which provides a partial answer.

* * * * *

In order to study the importance of collisions with interstellar matter we may then turn our attention to the relative abundance of the various

[7] A. H. Compton and I. A. Getting, *Phys. Rev.* **47,** 817 (1935).
[8] M. S. Vallarta, *Rev. Mod. Phys.* **21,** 356 (1949).
[9] Auger, Daudin, Denisse, and Daudin, *Comptes Rendus,* **228,** 1116 (1949).
[10] G. Lemaître, *Rev. Mod. Phys.* **21,** 366 (1949).

elements that are found to be present in the incident radiation. The most abundant elements in the primary cosmic radiation are hydrogen, helium, carbon, oxygen, nitrogen, magnesium, silicon, etc.,[11] which are also the most abundant elements in the known universe.[12] The elements between helium and carbon in the periodic system, lithium, beryllium, and boron, are elements of very low abundance, as compared with the neighboring elements. In meteorites (Goldschmidt) and in the solar atmosphere (Russell) their abundance is some 100,000 times smaller than the abundance of oxygen.[12] In interstellar space, according to Spitzer[12] the abundance of Li is at least 10^{-1} and the abundance of Be $\sim 10^{-4}$ times smaller than the abundance of Na, the cosmic abundance of which is 500 times smaller than that of oxygen. If the parallelism between the abundances in the primary cosmic radiation and the cosmic abundances still holds even approximately for these light elements, we may expect Li, Be, and B to be practically absent from the primary cosmic radiation *near their source region*. But this will no longer be true for the *incident* radiation, if nuclear collisions are important. As is known from recent emulsion work,[13] Li, Be, and B are quite frequently produced as "splinters" in collisions of fast nucleons with heavy nuclei, such as Ag and Br. High energy collisions between protons and the more abundant heavy cosmic-ray nuclei such as carbon and oxygen are also expected to result frequently not in complete dissociation of these nuclei into nucleons and alpha-particles, but to leave behind a residual fragment of charge $Z = 3$, 4, or 5. We shall show below that on the assumption of equilibrium between the acceleration of cosmic-ray nuclei of charge $Z \geqslant 6$ and their destruction in collisions with interstellar hydrogen we must expect that Li, Be, and B nuclei, produced in these collisions as fragments, must have an abundance comparable to that of all the heavier nuclei together.

* * * * *

[11] H. L. Bradt and B. Peters, *Phys. Rev.* **77**, 54 (1950).

[12] H. Brown, *Rev. Mod. Phys.* **21**, 625 (1949).

[13] D. H. Perkins, Intern. Cosmic Ray Conf. Como, Italy, 1949. We are indebted to Dr. Perkins for communicating to us details of his results before publication.

II. The charge spectrum of nuclei with $2 < Z \leqslant 14$ at $\lambda_{geom} = 30°$ below 20 g/cm² of air

(1) CHARGE DETERMINATION

In a previous paper[11] we have shown that at least up to charge $Z = 14$ primary nuclei are completely stripped of electrons before they enter the earth's magnetic field and therefore arrive at a geomagnetic latitude $\lambda = 30°$ with a kinetic energy of at least 3·5 BeV/nucleon. The energy loss which these particles suffer by ionization in the first 20 g/cm² of atmosphere is only a small fraction of this minimum energy. They will therefore enter a stack of photographic plates exposed at $\sim$ 90,000 ft with relativistic velocities. Neglecting the relativistic increase of ionization, their specific energy loss in the photographic emulsions is therefore a function of their charge only and is proportional to Z^2.

The specific energy loss in emulsions is measured in this work by the grain density along the track. By comparing the grain density along tracks of electron-positron pairs produced with widely differing angular divergence it has been shown that the relativistic increase of grain density for electrons in the energy range between 10 mc^2 and 10^4 mc^2 is less than 7%.

* * * * *

If a given particle makes a track which can just be perceived against the random background, another particle of 15 times this ionizing power will leave a track so dense that clogging of grains makes grain counting difficult. In order to extend the range of applicability of the grain counting method we have exposed a closely packed stack of plates in which highly sensitive emulsions alternate with emulsions of very low sensitivity. The ionization loss of a particle which traverses the stack can therefore be studied in both types of emulsion.

The sensitive plates consisted of Eastman NTB3 emulsions developed by the standard temperature method first suggested by Dilworth and Occhialini and Payne.[14] The insensitive plates consisted of Eastman

[14] Dilworth, Occhialini, and Payne, *Nature*, **162,** 102 (1948).

NTA emulsions developed by the same method but using 20 parts of water to one part of D-19 developer. This development procedure was chosen such as to make the heaviest countable tracks in the NTB3 emulsions just barely detectable in the NTA emulsion.[15] In Figure 1

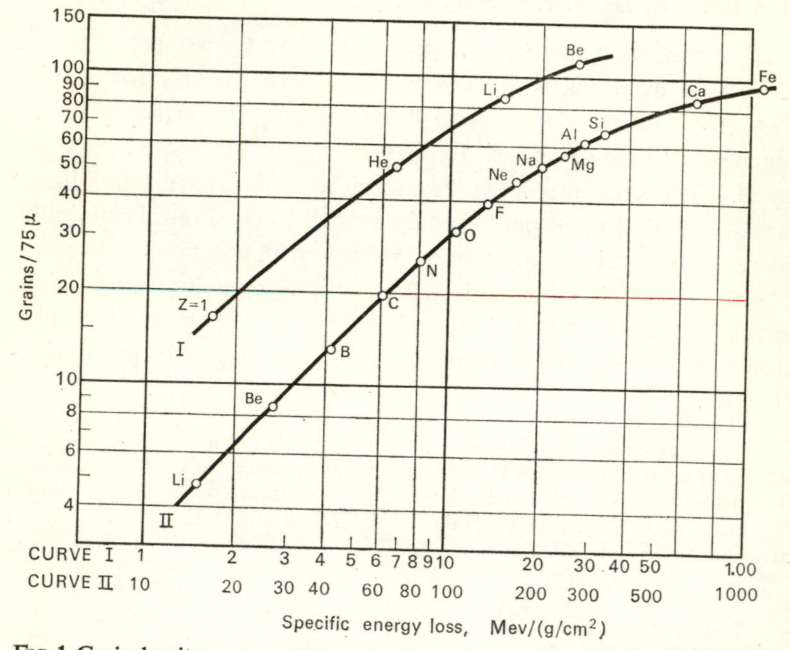

FIG. 1. Grain density *versus* specific energy loss. Curve I refers to NTB3 emulsion. Curve II refers to strongly underdeveloped NTA emulsions. The expected grain density of tracks due to various relativistic nuclei is indicated on the diagram.

the grain density for both emulsions is plotted against the specific energy loss of the particle.

Curve I refers to the electron sensitive emulsion. Shower particles and high energy electron–positron pairs created by γ-rays in the emulsion were used to establish the grain density corresponding to a relativistic singly charged particle. This point is marked $Z = 1$ on the curve. The points marked He, Li, Be giving the grain density corre-

[15] We are indebted to Mr. T. Putnam for carrying out the necessary development tests.

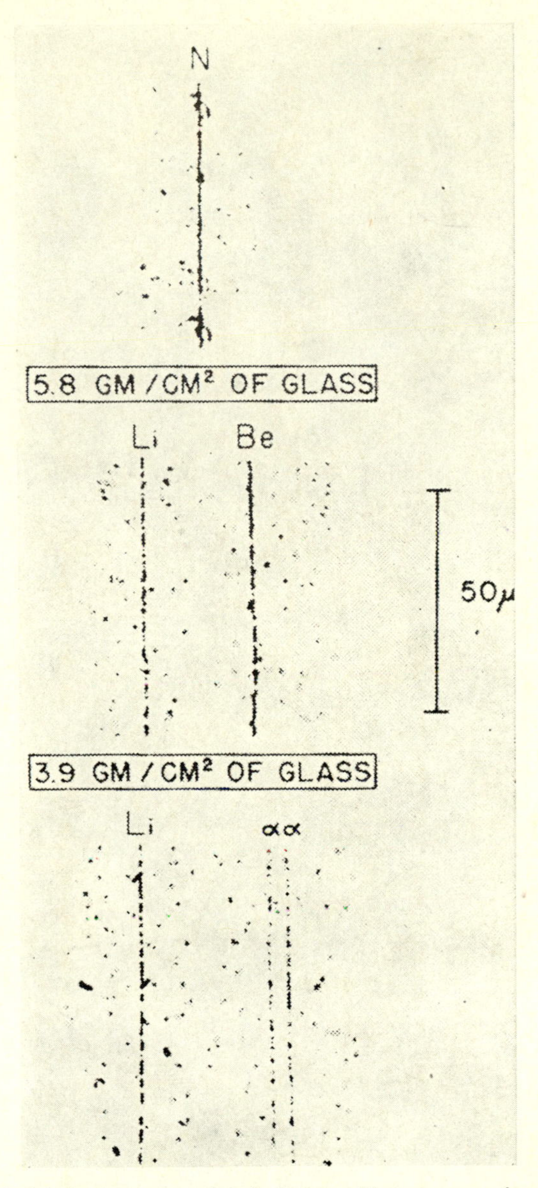

FIG. 2. A relativistic nitrogen nucleus is destroyed in two successive collisions as it traverses a stack of plates. The collisions occur in the glass between emulsions. Sections of the track are shown before and after the collisions. The nitrogen nucleus first disintegrates into a nucleus of lithium and a nucleus of beryllium, the latter then disintegrates into two α-particles.

sponding to 4, 9, and 16 times minimum ionization were determined from the particle tracks shown in Figure 2. Here a nitrogen nucleus after traversing several plates in the stack breaks up into a nucleus of lithium and a nucleus of beryllium, the latter in turn breaks up, after traversing several additional plates, into two α-particles. The small angle between the collision fragments and in particular the small divergence between the two α-particles from the disintegration of beryllium makes it certain that all the particles have relativistic energies. The grain density curve obtained in this way was checked by grain counting other relativistic α-particles and singly charged particles of long range.

Curve II refers to the emulsions which were strongly underdeveloped. Here the point marked Be was calibrated by using the relativistic beryllium nucleus previously identified and several other such nuclei found in the course of this investigation. The points marked O and Mg were fixed by grain counting tracks which in the neighboring sensitive emulsion were identified by δ-ray count as due to relativistic oxygen and magnesium nuclei. The calibration curve was then drawn through these three points and the grain density corresponding to relativistic nuclei of other charges was obtained from this curve by assuming a specific energy loss proportional to Z^2. A relativistic carbon nucleus is then expected to give a grain count of 20 grains/75 μ. Since previous work has established the existence of a strong carbon peak in the primary radiation, the appearance of such a peak in the charge spectrum (Figure 3) at the predicted value of grain density establishes the correctness of the calibration curve used for the underdeveloped emulsions.

* * * * *[a]

(3) THE CHARGE SPECTRUM OF PRIMARY NUCLEI AT $\lambda = 30°$

The tracks obtained in the surveys were grain counted in the underdeveloped (NTA) emulsions. Four tracks which in the sensitive emulsion indicated an ionization loss equal to 9 times minimum ionization

[a] Superscript letters refer to Notes at end of articles.

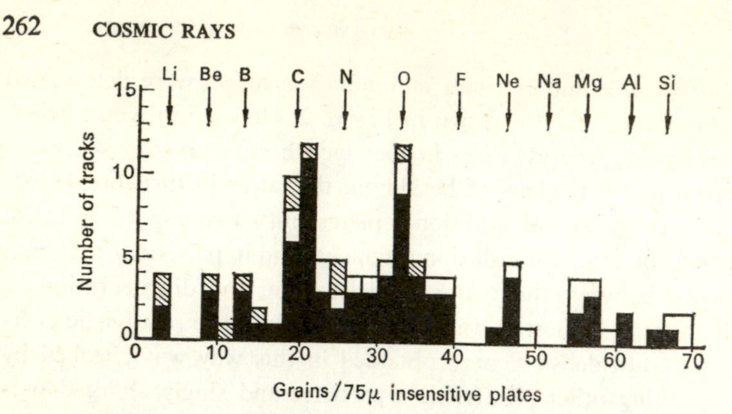

FIG. 3. Histogram of 111 tracks *vs* grain density in underdeveloped NTA emulsions. The solid squares represent a systematic survey. The open squares represent tracks obtained in a less systematic manner in the course of this work. Shaded squares represent relativistic fragments produced by heavier primary nuclei in collisions in the stack. The expected grain densities from different nuclei obtained from Fig. 1 are indicated by arrows. Lithium tracks as explained in the text are included in the diagram although they were identified in neighboring emulsions of higher sensitivity.

and are therefore ascribed to lithium nuclei did not leave a visible track in the NTA plates. For completeness sake they have been included, however, in the histogram Figure 3 at their calculated (but unobservable) grain density of 4·8 grains/75 μ. The grain density of all other tracks as determined in the NTA plates is shown in Figure 3. The charge values corresponding to these grain densities were obtained from Curve II, Figure 1 as explained above.

The solid squares in the histogram of Figure 3 refer to tracks obtained in a systematic survey. The open squares represent additional tracks obtained in a less systematic manner. The shaded squares represent relativistic fragments of heavier nuclei which suffered collisions in the glass.

Figure 3 confirms our previous result that among the nuclei with charge $Z > 2$ carbon and oxygen are the most abundant. They are responsible for the two most prominent peaks of about equal intensities. Though nitrogen nuclei have been identified, their number seems to be at least three times smaller than the number of either C or O nuclei and the nitrogen peak disappears in the high grain density "tail"

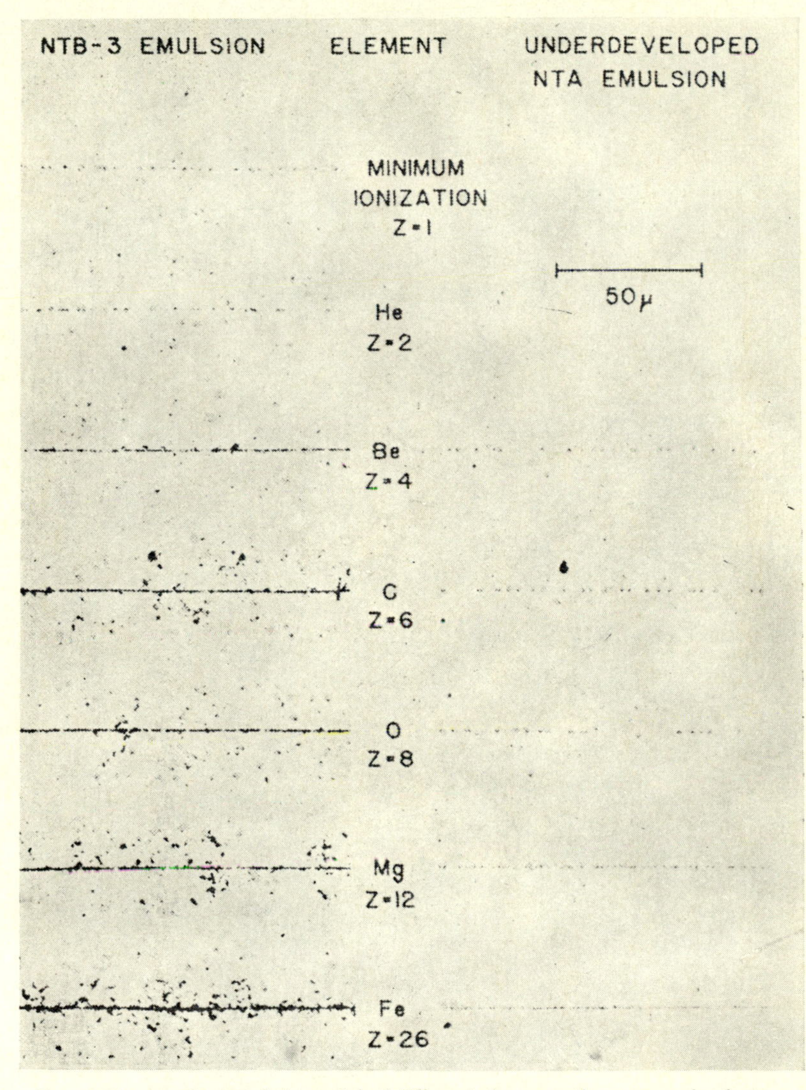

FIG. 4. Tracks of relativistic nuclei of different charge as they appear in electron sensitive NTB emulsions and in underdeveloped NTA emulsions.

of carbon. The "tail" of the oxygen peak makes it difficult to judge whether or not there is a significant number of fluorine nuclei. Between $Z = 8$ and $Z = 12$ we have a group corresponding to the charge $Z = 10$ (neon). Next comes the Mg, Al, Si group: these three elements are only poorly resolved, since the grain-density *vs.* energy-loss curve (Figure 1) flattens already considerably in this region.

* * * * *

Among the 93 tracks resulting from a systematic survey shown in Figure 3, 16 tracks belong to the Li, Be, B group. Six of those particles are fragments produced by heavier nuclei making collisions in the stack of plates. This is a very large fraction, in view of the fact that few of the particles of charge $Z > 5$ are fragments of still heavier nuclei produced in collisions in the stack.

The average length per track in the stack (from the point of entrance to either the point of exit or the collision) is about 15 g/cm^2, about two-thirds of the total amount of air above the balloon. The number of secondaries produced in the air overhead will therefore be somewhat larger than the number of secondaries produced in the glass and it is therefore reasonable to assume that also the remaining light nuclei (3 Be, 5 B, and 2 Li) that enter the stack from the outside are secondaries, fragments of heavier nuclei disintegrated in the air above the balloon. This indicates that at the top of the atmosphere the flux of particles of charge $3 \leqslant Z \leqslant 5$ is zero or very small. A more accurate evaluation of this flux is made in the next chapter.

The flux of nuclei with $6 \leqslant Z \leqslant 10$, extrapolated to the top of the atmosphere, had been determined previously:[11]

$I = (3{\cdot}5 \pm 0{\cdot}6) \times 10^{-4}$ nuclei/cm^2 sec sterad. for a geomagnetic latitude of 30°. The data of this flight, which was carried out near the same geomagnetic latitude, yield a flux value of $(3{\cdot}3 \pm 0{\cdot}7) \times 10^{-4}$ in good agreement with the previously determined value.

Excluding fragments produced in the glass we find in a systematic survey below 20 g/cm^2 of residual atmosphere:

	Li	Be	B	$6 \leqslant Z \leqslant 8$	$Z > 8$
No. of nuclei	2	3	5	51	19

At this altitude then the flux of Li, Be, and B nuclei constitutes approximately 20% of the flux of CNO nuclei.

III. The production of Li, Be and B fragments in energetic collisions between nuclei

Let us assume that Li, Be, and B nuclei are not found with any appreciable abundance among the true primaries; that is, among the nuclei originally accelerated at the source of the cosmic radiation. Li, Be, and B nuclei, observed at a certain depth h below the top of the atmosphere, may then be fragments produced in collisions of heavier cosmic-ray nuclei: (1) with air nuclei at the top of the atmosphere or (2) with interstellar matter (mostly hydrogen).

In order to estimate the contribution from these two sources we have to know the probability $p_A(Z)$ and $p_H(Z)$ that a stable fragment of Li, Be, or B will be produced in a high energy collision of a nucleus of charge Z with an air nucleus or a hydrogen nucleus respectively. For the purpose of this discussion we shall divide the heavy cosmic-ray nuclei into 3 groups:

A nuclei	$(3 \leqslant Z \leqslant 5)$,
B nuclei	$(6 \leqslant Z \leqslant 9)$,
C nuclei	$(10 \leqslant Z)$.

Below $h = 20$ g/cm^2 of residual atmosphere the relative number of these three components is approximately $N(A) : N(B) : N(C) = 1 : 6 : 2$.

(1) COLLISIONS IN AIR

(a) *The probability $p_A(Z)$ of producing an A nucleus in a collision of B or C nuclei with nuclei of the atmosphere*

The mean free path for collisions of B and C nuclei with nitrogen and oxygen nuclei of the atmosphere and the type of fragmentation occurring in such collisions may be inferred from the observed collision of such nuclei in the glass of photographic plates.

TABLE 1. COLLISION MEAN FREE PATH IN GLASS

Type of nuclei	Aggregate length of track observed (g/cm²)	Number of collisions observed	Mean free path (g/cm²)
A nuclei (Li, Be, B)	237	5	(47·4)
B nuclei (C, N, O)	2030	59	34·5 ± 4
C nuclei ($10 \leqslant Z \leqslant 18$)	954	36	26·5 ± 4
($19 \leqslant Z \leqslant 26$)	256	11	23·3 ± 7

Table 1 gives the collision mean free path in glass for *A*, *B*, and *C* nuclei. The values obtained for *B* and *C* nuclei are in agreement with those reported earlier.[11] The mean free path for *A* nuclei (λ_A) is based on an as yet insufficient number of collisions and is only given as an indication that, as expected, it is longer than λ_B or λ_C.

TABLE 2. STABLE RELATIVISTIC NUCLEI OF CHARGES $Z > 2$ RESULTING FROM COLLISIONS

Type of primary nuclei	Fragments							Number of collisions	Number of *A* nuclei produced per collision
	Li	Be	B	C	N	O	$Z > 8$		
B nuclei	6	6	3	6	3	—	—	65	0·23
C nuclei	5	4	—	8	3	3	9	38	0·24
	11	10	3	14	6	3	9	103	0·23

Table 2 above shows the number of stable relativistic nuclei of charge $Z > 2$ observed to result from 103 collisions of *B* and *C* nuclei in glass. Thus we get 0·23 *A* type nuclei (Li, Be, or B) produced per collisions of *B* or *C* type nuclei in glass. We may assume that the fragmentation of these nuclei in glass (mostly SiO_2) is very nearly identical to that occurring in air:

$$p_A(B) = p_A(C) = 0{\cdot}23$$

and that their collision mean free paths in air are about 10% less than the corresponding paths in glass.

(b) *Relativistic Li, Be, and B fragments produced in the air above the balloon*

We can now calculate the number of A nuclei which should be observed under 20 g/cm^2 of air.

* * * * *

Our results therefore indicate for the flux ratio (R) of A and B nuclei at the top of the atmosphere a value $0 \leqslant R < 0{\cdot}1$.[b]

We conclude that lithium, beryllium, and boron are rare elements in the region where cosmic-ray particles are accelerated.

(2) COLLISIONS WITH INTERSTELLAR HYDROGEN

(a) *The probability $p_H(B)$ and $p_H(C)$ of producing A nuclei in collision between B or C nuclei and protons*

Consider a C, N, or O nucleus at rest being struck by an incoming particle. If the incident particle is a heavy ion (say oxygen or silicon) the nuclear material of the target will disintegrate in a certain way and the probability of obtaining *from the target material* a fragment of charge $3 \leqslant Z \leqslant 5$ is $p_A(B) = 0{\cdot}23$, the same number which was obtained in the previous sections by observing collisions in which target and projectile were interchanged. If the stationary C, N, or O nucleus is not struck by a heavy ion but by a relativistic nucleon we would obtain a star with fewer prongs (less complete disintegration) and therefore a higher probability for a fragment of charge $3 \leqslant Z \leqslant 5$. This probability is the quantity $p_H(B)$ defined previously. A lower limit for $p_H(B)$ is therefore given by

$$p_H(B) > p_A(B) = 0{\cdot}23.$$

The same argument will apply for collisions of C nuclei ($Z \geqslant 10$) although we would expect that the probabilities depend less strongly on the weight of the nucleus with which the C nuclei collide. Therefore we have also

$$p_H(C) \geqslant p_A(C) = 0{\cdot}23.$$

A more accurate estimate of these probabilities can be made in the following way.

D. H. Perkins[13] has studied the production of heavy nuclear "splinters" in Ag and Br stars in nuclear emulsions exposed both at mountain altitude and in the stratosphere. He finds in all those collisions where at least half of the nuclear matter is disintegrated (number of prongs > 12) about 22% lead to the emission of a stable Li, Be, or B nucleus of more than 30 μ-range. The frequency of such fragments is found to be independent of the number of shower particles produced in the collision and therefore independent of energy at sufficiently high energies. From his data one obtains therefore $p_H(Z) \sim 0{\cdot}22$ (for $Z \sim 40$).

No direct measurements of $p_H(Z)$ are available for smaller values of Z, but for $Z = 6$ to 8 an estimate of this probability can be made from an analysis of stars produced in the gelatine of nuclear emulsions. Harding[16] has made such an analysis using thin gelatine layers imbedded in nuclear emulsions. (The emulsions used were not electron sensitive.)

He found (α) that the size frequency distribution of stars made in the gelatine (C, N, O nuclei) is essentially the same as that for stars of less than 6 prongs made in ordinary emulsions.

(β) That stars of more than 5 prongs are absent or very rare.

(γ) That $\sim 36\%$ of stars in ordinary emulsions must be due to the disintegration of C, N, or O nuclei.

(δ) That charge conservation requires that the number of fragments of charge $Z > 2$ from light nuclei is on the average about 0·7 per star.

In order to check whether the probability of producing a fragment of charge $Z > 2$ in a collision between a light nucleus and a nucleon

[16] J. B. Harding, *Nature*, **163**, 440 (1949).

is as high as indicated by Harding's data, even if one confines oneself to high energy primaries, we have made an analysis in an ordinary electron sensitive emulsion exposed at balloon altitudes of stars with less than 6 prongs produced by a singly charged relativistic primary particle. According to Salant *et al.*,[17] almost all such events represent disintegration of C, N, O nuclei contained in the gelatine of the emulsion.

Counting as prongs all tracks including recoils but excluding the incident primary we have examined 33 stars with a total of 131 prongs.

The nuclear fragments of an average star produced by a singly charged particle in the light nuclei of the emulsion must carry 7·95 units of charge, assuming a contribution from carbon, nitrogen and oxygen nuclei proportional to their geometric cross-sections (water content corresponding to 50% humidity included). The 33 stars with a total of 131 prongs found in the survey will then emit a total of $33 \times 7{\cdot}95 =$ 263 units of charge. Of the 131 prongs, 37 could be identified as due to singly charged particles, leaving 226 charges to be divided among the remaining 94 prongs. Assuming, as is indicated by the data of Perkins,[13] that stable nuclear fragments of charge 3, 4, and 5 occur with approximately equal probabilities, and making the extreme assumption that none of the unidentified prongs were due to singly charged particles, at least 19 prongs must be due to fragments of charge $Z > 2$ ($Z_{Av} = 4$) and the total charge is accounted for by

37 prongs caused by particles of $Z = 1$	37 charges
75 prongs caused by particles of $Z = 2$	150 charges
19 prongs caused by particles of $Z_{Av} = 4$	76 charges
131 prongs	*263*

Thus at least 0·6 A nuclei are produced in an energetic collision between a proton and a B nucleus, in agreement with Harding's result. It would be desirable to measure this quantity directly by studying stars produced by single charged relativistic primaries in gelatine layers

[17] Salant, Hornbostel, Fisk, and Smith, *Phys. Rev.* **79**, 184 (1950).

imbedded in electron sensitive emulsions. In the absence of such measurements we shall use as lower limit

$$p_H(B) = p_H(C) = 0{\cdot}23,$$

although present evidence indicates that this limit is considerably lower than the true value.

(b) *Production of relativistic Li, Be, and B nuclei as collision fragments in interstellar space*

* * * * *c

According to Perkins,[13] the velocity of the "splinters" in the rest system of the Ag or Br target nucleus rarely exceeds the value $\beta_0 = v_0/c = 0{\cdot}1$. The velocity of the Li, Be, and B fragments in two- or three-pronged stars produced by fast protons in the gelatine also in general cannot exceed this value, since otherwise these fragments would give rise to easily identifiable prongs. A relativistic *B* or *C* nucleus with total energy per nucleon (ε) will give rise to an *A* nucleus with total energy per nucleon (ε') where ε'/ε lies within the range of values given by $(1 \pm \beta\beta_0)/(1-\beta_0^2)$ and β is the velocity of the *B* or *C* nucleus. Since at geomagnetic latitude 30° $\beta \approx 1$ and since $\beta_0 \lesssim 0{\cdot}1$ we have $|\varepsilon - \varepsilon'|/\varepsilon \lesssim 10\%$. Thus at relativistic energies *A* nuclei produced in collisions in interstellar space will essentially have the same velocity spectrum as the *B* and *C* nuclei from which they derive.

(c) *The relation between the flux of A nuclei and nuclear collisions in interstellar space*

Let us assume that *B* and *C* nuclei after being accelerated at the source of cosmic radiation are retained by magnetic fields in some region surrounding the source until they are destroyed by collision with interstellar hydrogen gas. On these assumptions the radiation will contain a certain fraction of *A* nuclei produced in collisions even if no *A* nuclei were originally accelerated at the source.

The relative number of A and B nuclei which we would expect to be incident at the top of the atmosphere is then

$$R = p_H(B)(\sigma_B/\sigma_A)+p_H(C)(\sigma_C/\sigma_A)(N_C/N_B).$$

Here $p_H(B)$ and $p_H(C)$ are the probabilities that the collision between a B or C nucleus with hydrogen will lead to the formation of an A nucleus. $\sigma_{A,B,C}$ are the cross-sections for the disintegration of A, B, or C nuclei in collision with hydrogen and $N_C/N_B = \frac{1}{3}$ is the relative intensity of the C and B component. Assuming the cross-sections to be proportional to the geometric areas of the nuclei we obtain

$$R \geqslant 0{\cdot}5 \quad \text{if we use} \quad p_H(B) = p_H(C) \geqslant 0{\cdot}23.$$

If we use the more reasonable values discussed in section IIa:

$$p_H(B) > 0{\cdot}6, \quad p_H(C) > 0{\cdot}23 \quad \text{we obtain} \quad R > 1{\cdot}0.$$

Thus collision equilibrium of primary particles in hydrogen gas should lead to a composition in which the number of A nuclei is at least half as great as that of B nuclei and probably considerably larger.

The ratio $R = 0{\cdot}5$ is also obtained if we make the (less plausible) assumption that most of the interstellar collisions occur not with hydrogen nuclei but with dust particles (with an assumed average composition comparable to SiO_2).

Since the measurements discussed in section III(1), give a value for R less than 0·1 we conclude that only a small fraction of the observed primary cosmic-ray particles suffer nuclear collisions between the time they are accelerated and their arrival at the top of the atmosphere.

IV. On the relation between the chemical composition of primary radiation and theories on the origin of cosmic rays

In a previous paper[11] we have shown that if the primary cosmic radiation were in equilibrium with its collision products most of its proton and helium would have to be ascribed to collision fragments from heavier nuclei. In that case in order to account for the observed chemical composition of the radiation we would have to assume that the amount of hydrogen and helium accelerated at the source is at most equal in *weight* to the amount of heavier nuclei. Therefore

(α) either the source region where cosmic-ray particles are accelerated contains a very much smaller fraction of hydrogen and helium than those regions of the universe which are accessible to spectral analysis, or

(β) the acceleration of heavy particles ($Z > 2$) is greatly favored by the accelerating mechanism compared to the acceleration of light particles ($Z \leqslant 2$). Since an electromagnetic acceleration process requires that the atom is initially ionized one might expect that elements with low ionization potential are more frequently accelerated than those with high ionization potential. Such a selective action in the accelerating process does not seem to exist. For instance Brown[12] gives the relative abundance of the elements oxygen (ioniz. pot. 13·55 eV), neon (ioniz. pot. 21·47 eV), magnesium (ioniz. pot. 7·6 eV) as 5 : 1 : 2. A comparison of these ratios with the histogram of Figure 3 (observed ratio 22 : 5 : 5) shows that the ratios of these elements in the primary cosmic radiation do not differ significantly from the universal abundance ratios in spite of the large differences in ionization potentials. In view of this it seems difficult to think of an electromagnetic accelerating mechanism which strongly favors the acceleration of heavy elements.

These difficulties are removed if, based on the measurements described in the previous sections, we conclude that most of the cosmic-ray particles have *not* undergone nuclear collisions between their acceleration and their arrival. The chemical composition of primary particles will then reflect the composition of elements at the source; the relative number of protons and α-particles in the primary beam will not be increased compared to heavier elements by their longer mean free path for nuclear collisions or by the addition of collision fragments.

The absence or scarcity of lithium, beryllium, and boron nuclei in the primary radiation indicates therefore that as far as the distribution between light and heavy elements is concerned the chemical composition at the source is similar to the average chemical composition of the universe.

If we assume an electromagnetic accelerating process the chemical composition *within* the heavy component indicates that the source region differs from stellar atmospheres; in the source even the noble gas atoms are predominantly in an ionized state and therefore the relative abundance of ions in the source region parallels the relative abundance of elements.

The results derived in the previous sections can be expressed by the statement that for most of the primary particles of charge $Z > 2$ and kinetic energy per nucleon $\varepsilon > 3{\cdot}5$ BeV, the time τ between their acceleration and their entry into the earth's atmosphere must be such that $\tau \leqslant 10^6/\varrho$ years, where ϱ is the number of atoms per cc along their trajectory. This condition could be satisfied:

(α) If the particles came from outside the galaxy (where $\varrho < 10^{-3}$) as part of primordial matter. They do not, however, show the large deficiency in protons predicted by Lemaître.[10]

(β) If they were of galactic, but relatively recent, origin and did not have time to establish equilibrium with their collision products.

This possibility cannot as yet be disregarded, but recent work by Arnold and Libby[18] on the C^{14} content of historical relics seems to prove that the intensity of cosmic radiation has undergone no appreciable change at least during the last 20,000 years.

(γ) If they were produced in the galaxy but leak out at an appreciable rate.

(δ) If they were produced in the galaxy (in the neighborhood of stars) but adiabatically decelerated in interstellar space in a time short compared to τ.

This deceleration must be much more effective than the deceleration caused by ionization losses, since the ionization loss in hydrogen in a distance corresponding to a mean free path for nuclear collision is small for all particles whatever their charge.[d]

* * * * *

(ε) If cosmic-ray particles were absorbed by bodies larger than bricks (such that the particle is absorbed together with its fragments) and

[18] J. R. Arnold and W. F. Libby, *Science*, **110**, 678 (1949).

that the mean free path for collision with these larger objects is shorter than that for collision with nuclei.

Assuming the most frequently quoted value for the average number of atoms in interstellar space ($\varrho = 1$), the last three conditions when applied to our galaxy have one feature in common: In order to account for the observed energy density of cosmic radiation an accelerating mechanism must be devised efficient enough to permit the transformation of approximately one percent of the entire stellar energy production into cosmic radiation. The difficulty of conceiving a high efficiency accelerating process which was pointed out by Richtmyer and Teller[2] has been increased by a substantial factor.

Apart from raising this serious difficulty about galactic origin, the apparent absence of nuclear collisions in the life history of primary particles seems to lead to difficulties with the specific theories proposed by Spitzer[4] and by Fermi.[1]

In Spitzer's theory the particles are originally accelerated as dust gains by radiation pressure in the vicinity of a supernova. Since noble gases could not be present in the dust grains in any appreciable amount, their presence in the primary cosmic ray beam and particularly the large primary helium flux[11] requires that these nuclei are fragments produced in nuclear collisions of the heavier nuclei present in dust grains. We have shown previously that the large helium flux requires complete equilibrium between heavy primary nuclei and their collision products. In view of the absence of other expected collision products in the primary beam such an explanation is no longer possible.

In the theory proposed by Fermi[1] particles are accelerated by collisions in our galaxy with inhomogeneous moving magnetic fields. Fermi points out that while protons can be accelerated and regenerated by such a process, heavy nuclei will have to be injected continuously with energies above 1 BeV per nucleon before they can gain energy by these collisions and therefore require an additional accelerating mechanism.

If one now assumes that heavy cosmic ray nuclei either leak out of the galaxy or are adiabatically decelerated or absorbed in large bodies in a time shorter than $10^6/\varrho$ years one must assume similar effects for the proton component. By depriving the protons of their ability to make nuclear collisions with interstellar hydrogen one would deprive Fermi's theory of its regenerative mechanism. By reducing the lifetime of fast particles to approximately one million years as compared to the 60 million years assumed in Fermi's theory it may be more difficult to obtain agreement with the observed energy spectrum.

It seems perhaps most promising to consider the absence of nuclear collision products in the cosmic ray beam as an argument in favor of solar origin or at least in favor of a source region sufficiently close such that the chemical composition of the radiation still reflects the abundance ratios at the source.

In the theory of solar origin proposed by Richtmyer and Teller[2] and by Alfvén[3] a magnetic field is assumed to exist in the planetary system, weak enough to escape detection by direct measurements at the earth, yet strong enough to obtain isotropy and time independence especially for particles of very high energy. In its present form, however, this theory allocates to the cosmic radiation a volume (radius $\sim 10^{17}$ cm) which seems to be too large to avoid nuclear collisions of primary nuclei with intraplanetary matter unless condition (ε) is satisfied and an appreciable fraction of the radiation is absorbed in meteoric material.

We wish to express our gratitude to Dr. John Spence of Eastman Kodak Company for preparing special plates, to Miss K. Haile and Mr. R. Richard who carried out most of the scanning, to Mr. T. Putnam for the careful development tests and the microphotographs.

Notes

[a] A description of the search for tracks is omitted. Long (inclined) tracks with ionization $> 5I_{\min}$ were found in a sensitive emulsion and followed. Their penetrating power and constancy of ionization showed that they were relativistic nuclei of $Z > 3$, not non-relativistic αs or protons.

[b] The actual calculation is not reprinted. As noted in Part 1 of this book, the ratio R obtained is rather too low.

[c] It is estimated that the relative abundance of A nuclei is not greatly altered by decay of unstable fragments, although it is noted that some Be^{10}, with a half-life of $2{\cdot}5 \times 10^{6}$ years, will be formed.

[d] If relativistic.

15. The Nature of Cosmic Radio Emission and the Origin of Cosmic Rays[†]

V. L. GINZBURG[‡]

1. Introduction

In order to formulate a theory of the origin of cosmic radation (c.r.) it is obviously necessary to have a certain amount of information on the composition, energy spectrum and spatial distribution of the c.r. outside the Earth's atmosphere, in the solar system, in the Galaxy and between galaxies. However, up to very recent times only data (and very incomplete at that) on cosmic rays in the immediate vicinity of the Earth (at top of atmosphere) have been available. As a result of this, various theories of origin of the cosmic rays were founded on rather arbitrary hypotheses and even the fundamental question concerning the sources and mechanism of acceleration of c.r. particles remained obscure. One of the main aims of the present paper is to demonstrate that with the development of radioastronomy this state of affairs has radically changed. Thus, if one interprets the non-thermal cosmic radio emission as being a type of magnetic bremsstrahlung (radiation from electrons accelerated in interstellar magnetic fields) then radioastronomical data permit one to determine the spectrum and concentration of relativistic electrons of the primary c.r. in the Galaxy and beyond it. Thus the theory of origin of c.r. can rest on a reliable foundation; it now becomes a branch of astrophysics based on empirical data. The nature of cosmic radio emission will be considered in the present paper and a theory of the origin of c.r. based on radio-

† *Nuovo Cimento*, **3**, Ser. X, Suppl., 38–48 (1956).

‡ Academy of Sciences of the USSR, Moscow.

astronomical data will be described. (A more detailed account of the theory is given elsewhere.[1–3]) [The raised figures refer to the list of works at the end of this paper.]

2. The Magnetic Bremsstrahlung Nature of Cosmic Radio Emission

Cosmic radio waves emitted with a continuous spectrum consist of radiation from discrete sources and of "general" cosmic radio emission which varies relatively slowly with the galactic coordinates. The general radiation can in turn be resolved into metagalactic and galactic components. Finally the galactic radio emission may be divided into thermal and non-thermal components. The first of these is the thermal radio emission from interstellar gas and, similarly to the gas, it is located in the plane of the Galaxy. Naturally, the corresponding maximum effective temperature of the radiation, T_{eff}, cannot exceed the kinetic temperature of the gas and therefore cannot exceed $\sim$ 10,000°. However, in the direction of the Galaxy anticentre and pole the optical thickness of the gas is small and even for waves in the metre band the thermal radio emission intensity is weak ($T_{eff} \lesssim 1000°$); the same may also be said about the metagalactic radio emission. On the other hand the total intensity of cosmic radio emission of wavelength say of 16·3 m travelling from the galactic pole corresponds to an effective temperature $T_{eff} \sim 50{,}000°$. This signifies that there exists some type of non-thermal general galactic radio emission which depends weakly on the galactic coordinates and which compared with thermal and metagalactic radiation is of decisive importance in the metre band. From an analysis of the cosmic radio emission isophotes it has been concluded[4] that the non-thermal galactic radio emission comes from a quasispherical sub-system, a sort of corona which surrounds the stellar Galaxy. The radius of this sub-system is $R \sim 3\text{–}5 \times 10^{22}$ cm. Diffuse interstellar gas with a concentration of $n \lesssim 0{\cdot}1\ \text{cm}^{-3}$ possesses about the same spatial distribution.[5]

A convincing proof of the existence in the Galaxy of radio sources distributed in the manner described above is the discovery[6] of a similar "radio corona" around the nebula M31 in Andromeda which resembles our own Galaxy.

What is the nature of the non-thermal galactic radio emission? Even at the present time attempts are made to explain the radio emission as originating in a large number of hypothetical radio stars invisible in the optical range of the spectrum. This assumption, which for a number of reasons has always seemed to us to be improbable, becomes altogether untenable after the discovery that the discrete sources of cosmic radio waves are nebulae and not stars. There is therefore absolutely no justification in assuming the existence of an enormous number[4] of radio stars possessing some very queer properties and a very unusual spatial distribution. The "magnetic bremsstrahlung hypothesis" which associates non-thermal cosmic radio emission with the waves radiated by relativistic electrons moving in the interstellar magnetic field[1-3, 7-9] seems to be much satisfactory in this respect. It will be shown in the following that not only this hypothesis can be made to agree with the probable values of the interstellar magnetic fields and with the concentration of relativistic electrons, but that it is very fruitful for radioastronomy as a whole and for the theory of origin of c.r.

As is well known, in a magnetic field H an electron moves along a helical path with a cyclic frequency

$$\omega_H = \frac{eH}{mc} \cdot \frac{mc^2}{E}, \tag{1}$$

where E is the total electron energy. In the ultra-relativistic case, which is the only one of interest here, the electron radiates electromagnetic waves almost exclusively into a narrow cone whose angular aperture is $\theta \sim mc^2/E \ll 1$ in the direction of instantaneous velocity. Therefore, if the electron moves in circle, an observer at rest will "see" radiation pulses of duration

$$\Delta t \sim \frac{r\theta}{c}\left(\frac{mc^2}{E}\right)^2 \sim \left(\frac{mc}{eH}\right)\left(\frac{mc^2}{E}\right)^2,$$

where r is the radius of the orbit. If, however, the electron moves along a helix and the angle α between velocity and field is not too small* then

* The radiation will be of a different nature if the angle $\alpha \lesssim \theta \sim mc^2/E$; this case is of interest for the theory of solar sporadic radio emission[11].

$\Delta t \sim (mc/eH_\perp)(mc^2/E)^2$, where $H_\perp$ is the component of the magnetic field perpendicular to the direction of motion. The radiation spectrum will correspondingly consist of overtones of the frequency ω_H and practically may be considered as a continuous spectrum, the maximum being at a frequency $\omega_{max} \sim 1/\Delta t \sim (eH_\perp/mc)(E/mc^2)^2$. Calculations show[10] that the energy radiated by an electron per second in a frequency range $d\nu$ is $P(\nu, E)\,d\nu$, where

$$\left.\begin{aligned} P(\nu, E) \equiv P(\nu) = 2\pi P(\omega) &= 16\frac{e^3H_\perp}{mc^2}\,p\left(\frac{\omega}{\omega_m}\right) \\ &= 16\frac{e^2}{c}\,\omega_{H\perp}\left(\frac{\omega}{2\omega_{H\perp}}\right)^{\frac{1}{3}} Y(u), \\ \omega_{H\perp} = \frac{eH_\perp}{mc}\cdot\frac{mc^2}{E}, \quad & \omega_m = \frac{eH_\perp}{mc}\left(\frac{E}{mc^2}\right)^2, \\ u = \left(\frac{\omega}{2\omega_m}\right)^{\frac{2}{3}}, \quad & Y(u) = \frac{p\left(2u^{\frac{3}{2}}\right)}{u^{\frac{1}{2}}}. \end{aligned}\right\} \tag{2}$$

Values of the function $Y(u)$ are given in Table 1; in the limiting cases

$$\begin{gathered} \frac{\omega}{\omega_m} \ll 1: \quad Y = 0{\cdot}256, \qquad \frac{\omega}{\omega_m} \gg 1: \\ Y(u) = \frac{(2\pi)^{\frac{1}{2}}}{16}\,u^{\frac{1}{4}} \exp\left[-(4/3)u^{\frac{3}{2}}\right]. \end{gathered} \tag{3}$$

The function $p(\omega/\omega_m)$ has a maximum at $\omega = 0{\cdot}5\omega_m$, where it equals 0·1. Thus at the maximum

$$\left.\begin{aligned} P(\nu_{max}) &= 1{\cdot}6\,\frac{e^3H_\perp}{mc^2} = 2{\cdot}15\times 10^{-22}H_\perp \text{ erg/cycle} \\ \nu_{max} &= 0{\cdot}5\,\frac{\omega_m}{2\pi} = 1{\cdot}4\times 10^6 H_\perp\left(\frac{E}{mc^2}\right)^2 \text{ cycle/s.} \end{aligned}\right\}. \tag{4}$$

The intensity observed at the Earth will be

$$I_\nu = \frac{1}{4\pi}\int P(\nu, E)N_e(E, \mathbf{r})\,dE\,d\mathbf{r}, \tag{5}$$

where $N_e(E, \mathbf{r})$ is the differential electron spectrum at the point $\mathbf{r}$.

TABLE 1

u	$Y(u)$	u	$Y(u)$
0·0	0·256	1·8	0·0085
0·2	0·204	2·0	0·0050
0·4	0·156	2·2	0·0028
0·6	0·115	2·4	0·0015
0·8	0·081	2·6	0·0008
1·0	0·055	2·8	0·0004
1·2	0·036	3·0	0·0002
1·4	0·023	3·5	0·00004
1·6	0·014	4·0	0·000006

Integration is carried out along the ray of vision and it is assumed that, due to the random distribution of the field $\boldsymbol{H}$, the radiation on the average is isotropic (hence the factor $1/4\pi$ in (5)). This assumption means that the formulae are correct to a factor of the order of unity. Absorption and possible influence of the refraction index (which may be different from unity) have not been taken into account as they are of no consequence in the following exposition.[1]

If the electron spectrum does not change along a path R and has the form

$$N_e(E) = KE^{-\gamma} \tag{6}$$

then

$$\begin{aligned} I_\nu &= \frac{R}{4\pi}\int_0^\infty P(\nu, E)N_e(E)\,dE \\ &= \frac{3}{\pi}(2\pi)^{(1-\gamma)/2}\frac{e^3H_\perp}{mc^2}\left[\frac{2eH_\perp}{m^3c^5}\right]^{(\gamma-1)/2} U(\gamma)KR\nu^{(1-\gamma)/2} \qquad (7) \\ &\simeq 1{\cdot}3\times 10^{-22}(2{\cdot}8\times 10^8)^{(\gamma-1)/2}U(\gamma)KRH_\perp^{(\gamma+1)/2}\lambda^{(\gamma-1)/2}\ \text{erg/cm}^2 \\ &\qquad\qquad \text{cycle steradian,} \end{aligned}$$

where $U(\gamma) = \int_0^\infty Y(u)u^{(3\gamma-5)/4}\,du$ and, for example, for $\gamma = 1$, $5/3$, 2, 3 and 7 the function $U(\gamma)$ has the corresponding values 0·37, 0·163, 0·128, 0·087 and 0·153.[a]

[a] Superscript letters refer to Notes at end of articles.

Data on non-thermal galactic radio emission in the band 1·5 m $< \lambda <$ 17 m indicate that in a first approximation*

$$I_\nu \simeq a\frac{c}{\nu} = a\lambda, \quad a \sim 5\times10^{-21}\ \text{erg/cm}^3\ \text{cycle steradian}. \tag{8}$$

From (7) and (8) we obtain

$$\gamma \simeq 3, \quad I_\nu \simeq 3{\cdot}1\times10^{-15}H_\perp^2 KR\lambda \simeq 5\times10^{-21}\lambda \tag{9}$$

and hence $KH_\perp^2 R \sim 1{\cdot}6\times10^{-6}$. For the reasonably highest values of $R \sim 5\times10^{22}$ cm and $H_\perp \sim 10^{-5}$ oersted one obtains a minimum value $K_{min} \sim 3\times10^{-19} \simeq 10^5\ (\text{eV})^2/\text{cm}^3$.[b]

To be cautious we put $K \simeq 10^6$ and thus arrive at the following electron spectrum:[b]

$$N_e(E) \simeq \frac{10^6}{E^3}(\text{eV})^{-1}\ \text{cm}^{-3},\ N_e(E > E_0) = \int_{E_0}^{\infty} N(E)\, dE \simeq \frac{10^6}{2E_0^2}\ \text{cm}^{-3}, \tag{10}$$

where E and E_0 are measured in eV. Hence $N_e(E > 10^9\ \text{eV}) \simeq 5\times10^{-13}$ which is not in contradiction with data on cosmic ray electrons near the Earth (for more details see elsewhere[9]).[c] On the other hand $N_e(E > 10^8\ \text{eV}) \simeq 5\times10^{-11}$, which in order of magnitude equals the concentration of all primary cosmic-ray particles incident on the Earth (due to high latitude cut-off $cp \geqslant 10^9$ eV). Taking into consideration the form of the function $Y(u)$, it may be shown that the spectrum (10) should refer to the region $2\times10^8 < E < 3\times10^9$; uncertainties in the data permit one to lower the value $N_e(E > 10^9\ \text{eV})$ by several times.

Therefore, as the magnitude of the interstellar magnetic field ($\sim 10^{-5}$ oersted) required by the hypothesis of the magnetic bremsstrahlung agrees with other estimates of this quantity and this hypothesis also yields a reasonable value for the electron concentration, one may conclude that the magnetic bremsstrahlung explanation of the non-thermal general galactic radiation is probably correct. Moreover, this hypothesis automatically explains the fact that not more than 1%

* More accurately $I_\nu = a\nu^{-\alpha}$, where $\alpha = 0{\cdot}8$–$1{\cdot}2$.

of the primary c.r. with $E \geqslant 10^9$ eV incident on the Earth consists of electrons. The reason of this is that electrons lose energy in magnetic bremsstrahlung (which is negligible for protons and nuclei). If one also assumes (and this seems quite natural) that electrons, protons and nuclei in the energy region $E \lesssim 10^{11}$ eV are generated by the primary c.r. sources with a spectrum of the same type of (6), with $\gamma = \gamma_0 \simeq 2$, then, due to deceleration of the electrons in the magnetic field, the electron spectrum will have the same form as (6), but with $\gamma = \gamma_0 + 1 \simeq 3$.[1-3, 9] According to (10) this is just the type of spectrum one would expect on the basis of radioastronomical data.*

The proposed magnetic bremsstrahlung mechanism may also be responsible for radio emission from powerful discrete sources.[7, 8] As a matter of fact in such sources as α Cassiopeiae, α Cygni, α Tauri, $T_{eff} \gtrsim 10^7$ degrees, whereas the electron kinetic temperature is much smaller. It is impossible to explain the radiation from such galactic sources as α Cassiopeiae or α Tauri as originating from hypothetical radio stars; the magnetic bremsstrahlung hypothesis, on the other hand, does not require any impossible assumption about the magnetic fields or the electron concentration in the sources.

If the electron spectrum and the magnetic field in the source are independent of the coordinates, then the radiation flux from the source will be

$$F_\nu = I_\nu \, d\Omega = \frac{V}{4\pi R^2} \int P(\nu, E) N_e(E) \, dE, \tag{11}$$

where V is the volume of the source and R is the distance from it. Under the optimum conditions (4)

$$F_\nu = 1 \cdot 6 \frac{e^3 H_\perp}{mc^2} \cdot \frac{N_e V}{4\pi R^2} = 1 \cdot 7 \times 10^{-23} H_\perp \frac{N_e V}{R^2} \text{ erg/cm}^2 \text{ cycle}, \tag{12}$$

where N_e is the concentration of electrons; it is considered[4] that these

* Wandering of protons and nuclei in interstellar space should not change their spectrum and therefore, near the Earth, $\gamma = \gamma_0 \simeq 2$, which is the experimentally observed value for $E \sim 10^9$–10^{11} eV.

possess an energy corresponding to the frequency $\nu = \nu_{max}$. For α Tauri: $R \sim 5 \times 10^{21}$ cm, $V \simeq (4\pi r^3/3) \sim 10^{56}$ cm^3, while in the metre band

$$F_\nu \cong 2 \times 10^{-20} \text{ erg/cm}^2 \text{ cycle}$$

and is independent on the frequency. Putting $H_\perp \sim H \sim 10^{-4}$ which seems to be allowed[12] we obtain from (12) $N_e V \sim 10^{50}$ and $N_e \sim 10^{-6}$ cm^{-3}; obviously, these values are the minimal ones. By using formulae (7) and (10) one can determine the electron spectrum, and the energy in discrete sources can be obtained. For α Tauri (Crab nebula) $\gamma \simeq 1$, $N_e V \sim 10^{51}$ for $E > 2{\cdot}5 \times 10^8$ eV, and the total energy of relativistic electrons with $E > 2{\cdot}5 \times 10^8$ is approximately $W \sim 10^{47}$–10^{49} erg (see ref. 12 for further details). Compared with other powerful sources the source in Taurus should be considered exceptional* because in the majority of powerful sources (mostly extragalactic) $\gamma \simeq 3$; for Cassiopeia one also has $N_e \sim 10^{51}$ and $W \sim 10^{48}$ erg. The Crab nebula (α Tauri) is undoubtedly an expanding envelope of the supernova of 1054 A.D. According to Šklovskij[14] all other powerful galactic discrete sources are also envelopes of supernovae (thus Cassiopeia is a supernova of 369 A.D.).

We thus arrive at the conclusion that the total energy of relativistic electrons in the envelopes of supernovae is of the same magnitude as the energy of the optical radiation produced during the outburst. The radiation of powerful extragalactic discrete sources, according to the mass of collected data, should also be attributed to magnetic bremsstrahlung.[15] Thus, radioastronomy presents the possibility of detecting relativistic electrons in various parts of the universe. In particular, it is found that large amounts of electrons exist in the envelopes of supernovae and, possibly, of novae.[12, 16, 17] It seems difficult to overestimate the significance of this circumstance for the theory of the origin of c.r.

* The large amount of high energy electrons in α Tauri makes it probable that the magnetic bremsstrahlung mechanism is also responsible for the continuous optical radiation of the nebula.[13]

3. The Origin of Cosmic Rays

Data on the distribution of relativistic electrons in the Galaxy speak against the theory of solar origin of c.r. and also against the hypothesis of their metagalactic origin. One must therefore conclude that the main part of c.r. is generated somewhere in our Galaxy; these rays fill the Galaxy and form a quasi-spherical system (mentioned in § 1) with $R \lesssim 5 \times 10^{22}$ cm. It is furthermore natural to suppose that the primary c.r. sources are supernovae and novae[1–3, 12, 17] the envelopes of which contain fast electrons and probably protons and nuclei. As time passes (in about 1000–3000 years) the supernovae envelopes smear out and the c.r. which previously diffused from the envelope become completely distributed throughout interstellar space. Now the particles wander through interstellar magnetic fields, diffuse towards the periphery of the Galaxy and simultaneously lose energy in nuclear collisions and, later on, in ionization collisions. Instead of nuclear collisions, electrons undergo bremsstrahlung collisions with the nuclei and with the electrons of the interstellar gas.

The electrons also lose energy in magnetic-bremsstrahlung processes discussed in § 1. For a number of reasons, the acceleration of c.r. particles in the interstellar space[18] does not seem to be an effective process and even at very high energies ($E \gtrsim 10^{13}$ eV/nucleon) it probably is of no significance (see Ginzburg[1]). This conclusion[2] is not affected by attempts, which seem very unconvincing to us, to retain the theory of interstellar acceleration by assuming that the lifetime of the particles in the Galaxy is determined by their rate of exit from it.[19, 20] On the other hand Fermi's statistical acceleration mechanism[18] should be very effective[21] in turbulent envelopes of supernovae, where it is probably the main acceleration mechanism. Such is the picture one can draw of the origin of c.r. using radioastronomy data. It is impossible for us in this paper to consider the theory in detail[1–3] and we will therefore discuss some of the more important points of the problem.

A most important condition which any theory of origin of c.r. must satisfy follows from energy considerations. If the cross-section for energy transfers between protons colliding in interstellar space is

$\sigma \simeq 2{\cdot}5\times10^{-26}$ cm^2, then on the average the lifetime of fast protons in the Galaxy should be $T \sim 4\times10^8$ years $\sim 10^{16}$ s (the concentration of interstellar hydrogen is taken to be on the average $n \sim 0{\cdot}1$ cm^{-3}). The energy density of cosmic rays is $w \sim 1$ eV/cm^3, the total energy being $W \sim wV \sim 10^{68}$ eV, as the volume of the Galaxy $V \sim (4\pi/3)R^3 \sim 10^{68}$ cm^3 for $R \sim 3\times10^{22}$ cm. It is thus clear that the power of the sources of primary c.r. should be $W/T \sim 10^{52}$ eV/s $\simeq 10^{40}$ erg/s. This value is probably too high by one or two orders of magnitude, as most likely the accepted values of W and R are too high. Thus

$$W/T \sim 10^{38}\text{–}10^{40} \text{ erg/s.} \tag{13}$$

Assuming that during supernova outbursts, say, 10^{48} erg appear as c.r. (see above) one finds that average power of the source should be $\lesssim 10^{38}$ erg/s as on the average supernovae outbursts take place at least once in 300 years $\sim 10^{10}$ s. It is likely that supernovae outbursts occur much more often, but because of absorption they remain invisible (this argument belongs to I. S. Šklovskij). Furthermore, according to the estimates in Ginzburg,[1,2] the contribution from novae may reach 3×10^{40} erg/s. Thus supernovae and novae can satisfactorily provide the energy balance. It should be noted that energy considerations suggesting supernovae as possible sources of c.r. were proposed previously.[22] However, these arguments acquire new significance when one associates them with radioastronomical data which confirm the generation of c.r. in the envelopes of supernovae. It should also be noted that on the average the power of the c.r. generated by the Sun is of $\sim 10^{21}$ erg/s and therefore even 10^{11} stars of the solar type would yield only 10^{32} erg/s, which is by 6–8 orders lower than the required value (see Šklovskij[13]).

It has already been mentioned that in envelopes of supernovae and novae the particles are mainly accelerated by a statistical mechanism. This mechanism is capable of accelerating protons and light nuclei* to the highest energies observed in cosmic rays, i.e. up to $E \sim 10^{18}$ eV.[1,21] At the same time it should be emphasized that many details of the

* In the simplest case of statistical acceleration a particle of energy Mc^2 acquires, when accelerated for a time t, energy $E = Mc^2 \exp[\alpha t]$, that is, the energy is proportional to the rest mass.

acceleration mechanism in expanding turbulent envelopes are still obscure. For this reason a reliable theoretical prediction of the spectrum of generated particles cannot be given; moreover for different sources the spectra are different (for details see Ginzburg[1]). The spectra of electrons, protons and nuclei generated in the envelopes are probably of the same form and on the average, for energies $E \lesssim 10^{11}$, $\gamma \simeq 2$. Magnetic bremsstrahlung losses in interstellar space result in a softening of the electron spectrum (and therefore $\gamma \simeq 3$; see above).

The proton and nuclear spectra will remain unchanged (even when nuclear collisions are taken into account) if the energy of the secondary products (i.e. energy after the collision) is

$$E_2 = \delta E_1, \tag{14}$$

where δ is independent or depends only slightly on the primary particle energy E_1. The explanation of this is that relation (14) is a scale transformation, and collisions of this type do not lead to changes in the power exponent in spectra of the type (6). There are data which support this point of view.[2, 23] Furthermore, according to the theory outlined here, one should expect that in a first approximation a stationary state should exist in the Galaxy, i.e. that the concentration of the various particles should be independent of time. On the other hand, the primary c.r. sources form a plane subsystem, lying in the plane of the Galaxy (this is exactly the way novae and probably supernovae are distributed). This leads us to the result that an approximate equilibrium number of Li, Be and B nuclei should be present in the c.r. incident on the earth; thus the concentration of these nuclei should be about four-tenths the concentration of all other nuclei.[3] This is exactly the value which the latest available data yield.[24] Furthermore, according to our point of view, the high latitude cut-off in the c.r. spectrum cannot be explained by ionization losses and is probably due to the existence of a magnetic moment of the solar system.[9] This conclusion is in accord with data[25] indicating that the high-latitude cut-off is of a magnetic nature.

Finally, special attention should be paid to secondary electrons and positrons produced in the Galaxy during nuclear collisions between cosmic protons and nuclei. Thus, basing on the data of Vernov *et al.*[23]

it should be expected that in each collision about 5% of the energy of the primary particles should be given to electrons and positrons.[2,3]

It can be shown[2] that in the absence of interstellar magnetic fields the amount of primary c.r. electrons and positrons with $E > 10^9$ eV should be comprised within $\sim$ 1–5%, even if these particles were produced exclusively in nuclear collisions. This value becomes about 10 times smaller if magnetic bremsstrahlung energy losses are taken into account. However, there are reasons to believe that electrons are also produced in the primary sources (see § 1), their number being larger than that of the secondary electrons. Similar considerations and also the computations carried out in § 1 indicate (if no other factors must be taken into account) that the number of primary c.r. electrons near the earth with $E \geqslant 10^9$ eV should be about 0·1–0·5% of the number of primary protons with $E \geqslant 10^9$ eV. Some of these electrons are secondary and can be separated from the primaries, as the number of secondary electrons and positrons should be about equal, while in supernovae envelopes one should expect that only electrons be accelerated. Accordingly, a study of the primary electron–positron component near the Earth seems to be of special importance. Other possible means of checking the ideas outlined in this paper are discussed elsewhere.[3] Therefore, in conclusion we confine ourselves to the statement that, as the previous discussion seems to indicate, the present data and conceptions support the picture of the origin of c.r. proposed in the paper and, at any rate, do not contradict it.

[1] V. L. Ginzburg, *Usp. Fiz. Nauk*, **51**, 343 (1953); *Fortschritte der Physik (Berlin)*, **1**, 659 (1954).

[2] V. L. Ginzburg, *Dokl. Akad. Nauk SSSR*, **99**, 703 (1954).

[3] V. L.Ginzburg, *Proceedings of 5th Conference on Cosmogony* (Moscow, 1955).

[4] I. S. Šklovskij, *Astron. Žu. SSSR*, **29**, 418 (1952).

[5] S. B. Pikel'ner, *Dokl. Akad. Nauk SSSR*, **88**, 229 (1953).

[6] J. E. Baldwin, *Nature*, **174**, 320 (1954).

[7] H. Alfvén and N. Herlofson, *Phys. Rev.* **78**, 616 (1950); K. O. Kiepenheuer, *ibid.* **79**, 738 (1950).

[8] V. L. Ginzburg, *Dokl. Akad. Nauk SSSR*, **76**, 377 (1951); G. G. Getmancev, *ibid.* **83**, 557 (1952).

[9] V. L. Ginzburg, G. G. Getmancev and M. I. Fradkin, *Proceedings of 3rd Conference on Cosmogony*, p. 149 (Moscow, 1954).

[10] V. V. Vladimirskij, *Zu. Éksper. Teor. Fiz.* **18**, 393 (1948).

[11] G. G. Getmancev and V. L. Ginzburg, *Dokl. Akad. Nauk SSSR*, **87**, 187 (1952).
[12] V. L. Ginzburg, *Dokl. Akad. Nauk SSSR*, **92**, 1133 (1953).
[13] I. S. Šklovskij, *Dokl. Akad. Nauk SSSR*, **90**, 983 (1953).
[14] I. S. Šklovskij, *Astron. Žu. SSSR*, **30**, 15 (1953).
[15] I. S. Šklovskij, *Dokl. Akad. Nauk SSSR*, **98**, 353 (1953).
[16] J. G. Bolton, G. J. Stanley and O. B. Slee, *Austral. J. Phys.* **7**, 110 (1954).
[17] I. S. Šklovskij, *Dokl. Akad. Nauk SSSR*, **91**, 475 (1953).
[18] E. Fermi, *Phys. Rev.* **75**, 1169 (1949).
[19] E. Fermi, *Astrophys. J.* **119**, 1 (1954).
[20] P. Morrison, S. Olbert and B. Rossi, *Phys. Rev.* **94**, 440 (1954).
[21] V. L. Ginzburg, *Dokl. Akad. Nauk SSSR*, **92**, 727 (1953).
[22] D. ter Haar, *Rev. Mod. Phys.* **22**, 119 (1950).
[23] S. N. Vernov, N. L. Grigorov, G. T. Zacepin and A. E. Čudakov, *Izv. Akad. Nauk SSSR, Ser. Fiz.* **19**, 493 (1955).
[24] M. F. Kaplon, J. H. Noon and G. W. Racette, *Phys. Rev.* **96**, 1408 (1954).
[25] R. A. Ellis, M. B. Gottlieb and J. A. Van Allen, *Phys. Rev.* **95**, 147 (1954).

Notes

[a] And $U(2{\cdot}4) = 0{\cdot}105$.

[b] Figures misprinted in the original text have been corrected here.

[c] For comparison with the experimental data, note that the flux I is related to the particle density by the relation $I = cN/4\pi$: thus 5×10^{-13} cm^{-3} corresponds to $1{\cdot}19\times10^{-3}$ particles cm^{-2} s^{-1} sr^{-1} above this energy threshold.

16. Cosmic-ray Modulation by Solar Wind[†*]

E. N. Parker[‡]

It is shown that the hydrodynamic outflow of gas from the Sun observed by Biermann results in a reduction of the cosmic-ray intensity in the inner solar system during the years of solar activity. The computed cosmic-ray energy spectrum so closely resembles the observed spectrum at Earth that we suggest the outflow of gas to be the explanation for the 11-year variation of cosmic-ray intensity.

It is also suggested that perhaps the Forbush-type decrease, which is a local geocentric phenomenon, is the result of disordering of the outer geomagnetic field by the outflowing gas from the Sun.

I. Introduction

In a recent paper[1] it was shown that the hydrodynamic flow[2] of gas outward in all directions from the Sun, as observed by Biermann,[3] stretches out the magnetic lines of force of the solar magnetic fields, and leads to an essentially radial magnetic field in the inner solar system; the field density at the point (r, θ, φ), well removed from the Sun is

$$B(r, \theta, \phi) \simeq B(a, \theta, \phi)(a/r)^2,$$

† *Phys. Rev.* **110,** 1445–1449 (1958). Section III, on Geocentric modulation, is omitted.

* Assisted in part by the Office of Scientific Research and the Geophysics Research Directorate, Air Force Cambridge Research Center, Air Research and Development Command, U.S. Air Force.

‡ Enrico Fermi Institute for Nuclear Studies, University of Chicago.

[1] E. N. Parker, *Phys. Rev.* **109,** 1874 (1958).

[2] E. N. Parker, *Astrophys. J.* (to be published).

[3] L. Biermann, *Z. Astrophys.* **29,** 274 (1951); *Z. Naturforsch.* **7a,** 127 (1952); *Observatory,* **107,** 109 (1957).

where a is the radius (7×10^5 km) of the Sun, r is distance from the centre of the Sun, and θ and φ are the usual polar and azimuthal angles. A field density of 1 gauss at the Sun[4] yields about 2×10^{-5} gauss at the orbit of the Earth ($r = 1{\cdot}5\times10^8$ km).

The high velocity (500–1500 km/sec) and low density ($\sim$ 500 ions/cm^3) of this outward streaming gas, the solar wind, leads to anisotropic thermal motions as a consequence of the anisotropic expansion. And it was pointed out[1] that when the gas pressure in the direction parallel to the magnetic field exceeded the pressure perpendicular by an amount Δp which was greater than $B^2/4\pi$, the dynamical equations[5] for the plasma predict instability of the magnetic field to transverse perturbations. It was estimated that the instability should set in somewhere in the vicinity of the orbit of Earth, thereby leading us to the conclusion that the inner solar system is surrounded by a thick shell of disordered magnetic field of perhaps 2×10^{-5} gauss.

The existence of an enclosing shell of disordered field of $\sim 10^{-5}$ gauss had been inferred earlier from considerations[6] of the decay of the cosmic-ray intensity from the solar flare of February 22nd, 1956, which suggested that the shell had (at that time) a thickness of about 4 astronomical units, and extended from about the orbit of Mars to the orbit of Jupiter.

Such a heliocentric shell of disordered field will, in the presence of the 500–1500 km/sec solar wind, have significant effects on the cosmic-ray intensity in the inner solar system.[7] Each irregularity of the magnetic field in the disordered shell is carried outward in the solar wind. The cosmic rays will tend to be swept ahead of the field fluctuations. Equilibrium is reached when the rate at which cosmic rays can diffuse in through the shell is equal to the rate at which they are removed by convection. Hence the cosmic-ray intensity inside the shell is less than outside.

In addition to this heliocentric reduction of the cosmic-ray intensity

[4] H. W. Babcock and H. D. Babcock, *Astrophys. J.* **121,** 349 (1955).

[5] E. N. Parker, *Phys. Rev.* **107,** 924 (1957).

[6] Meyer, Parker, and Simpson, *Phys. Rev.* **104,** 768 (1956).

[7] E. N. Parker, *Phys. Rev.* **103,** 1518 (1956).

there may be a further effect of the solar wind on the cosmic-ray intensity. Polar magnetic stations show that the magnetic lines of force extending beyond about 6 earth's radii (those lines of force which come down within about 25° of the magnetic pole) are continually agitated by the solar wind. And we may infer from this that the lines of force are always slightly disordered beyond 6 radii; and when the solar wind is high, the disordering is probably substantial. Such disordering of the outer geomagnetic field affects the entrance of cosmic rays, and thereby causes an observer on the surface of Earth to see a geocentric change in intensity.

II. Heliocentric modulation

It has been pointed out elsewhere[1] that we can expect disordering of the magnetic field in the heliocentric shell with scales up to a maximum of about $l = 2\times 10^6$ km; the proportionately slower growth of larger scales does not allow their full development in the time available.

The radius of curvature $P(\eta)$ of a cosmic-ray particle with rest energy Mc^2 and kinetic energy ηMc^2 is

$$P(\eta) = Mc^2[\eta(\eta+2)]^{\frac{1}{2}}/ZeB. \tag{1}$$

Hence the path of a 1-BeV proton in the shell field of $\sim 2\times 10^{-5}$ gauss has a radius of curvature of about $2{\cdot}6\times 10^6$ km, which we note is larger than the scale l of the disordering of the field in the heliocentric shell. Thus upon passing through the shell, a cosmic-ray proton undergoes a large number of random angular deflections of the order of $l/P(\eta)$. The largest deflection expected for a 1-BeV proton is, accordingly, of the order of $2/2{\cdot}6$ radians or about 45°; the smaller-scale ($< 2\times 10^6$ km) magnetic irregularities, and higher-energy particles, will result in smaller deflections. But the cumulative effect of n small angular deflections l/P is

$$\Theta(\eta) \simeq n^{\frac{1}{2}} l/P(\eta).$$

Therefore, if we define a collision to be a total deflection of $\Theta(\eta) = \pi/2$, then the mean free path is

$$L(\eta) = \pi^2 P^2(\eta)/4l \tag{2}$$

for a particle with energy η.

Elementary kinetic theory tells us that the coefficient of diffusion of particles with a velocity w and mean free path L is

$$\varkappa(\eta) \simeq \frac{1}{3}\omega(\eta)L(\eta)\ \text{cm}^2/\text{sec} = \frac{\pi^2}{12}\left(\frac{M^2c^5}{Z^2e^2B^2l}\right)\frac{[\eta(\eta+2)]^{\frac{3}{2}}}{(\eta+1)}. \tag{3}$$

If the diffusing medium (the magnetic irregularities) has a general motion $\mathbf{v}$, the diffusive and convective transport equation is[7]

$$\partial j(\eta)/\partial t = -\nabla.\quad [\mathbf{v}j(\eta)]+\nabla.\quad [\varkappa(\eta)\nabla j(\eta)] \tag{4}$$

for the density $j(\eta)$ of cosmic-ray particles with energy η.

Supposing that the disordered shell extends uniformly, and with spherical symmetry, from a solar distance $r = r_1$ out to $r = r_2$, we have upon integration of (4) that the steady-state cosmic-ray density $j_0(\eta)$ inside the shell is related to the galactic density $j_\infty(\eta)$ outside by

$$j_0(\eta) = j_\infty(\eta)\exp\left\{-\frac{12v(r_2-r_1)lZ^2e^2B^2(\eta+1)}{\pi^2M^2c^5[\eta(\eta+2)]^{\frac{3}{2}}}\right\}. \tag{5}$$

If we suppose that we have the typical values mentioned above, $l \approx 2\times10^6$, $B \approx 2\times10^{-5}$ gauss, $r_2-r_1 = 4$ astronomical units (6×10^{13} cm), and $v = 10^3$ km/sec, then we have

$$j_0(\eta) = j_\infty(\eta)\exp\left\{-2{\cdot}16(\eta+1)/[\eta(\eta+2)]^{\frac{3}{2}}\right\} \tag{6}$$

for protons. $j_0(\eta)/j_\infty(\eta)$ is plotted in Figure 1. Note that the result is a depression of the cosmic-ray spectrum at all energies; but particularly at low energies, where a sharp cutoff occurs. The mean particle energy inside the shell is increased as a consequence of the decreased intensity; the low-energy cosmic-ray particles are kept out or the inner solar system by the solar wind in the heliocentric shell.

But note that this is just the sort of depression of the cosmic-ray intensity that one observes during the 11-year cycle of solar activity. When

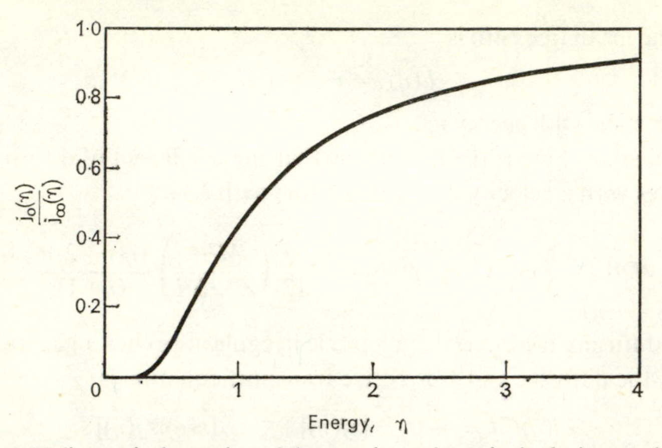

FIG. 1. Heliocentric depression of the cosmic-ray intensity in the inner solar system by the solar wind in the heliocentric shell of disordered magnetic field.

solar activity is at a minimum (as it was in 1954), the differential energy spectrum of the primary cosmic rays is observed[8] to be approximately of the form $j(\eta) \propto \eta^{-n}$, with $n \simeq 2{\cdot}7$ above about 2 BeV/ nucleon; below 2 BeV/nucleon n has a somewhat smaller value, but there is no indication[9] that $j(\eta)$ reaches a maximum before zero energy.

On the other hand, with the reappearance of solar activity following the minimum, the cosmic-ray intensity begins to decrease. Meyer and Simpson[8] have shown that the cosmic-ray intensity at 2 BeV was depressed by 50% during 1948 when the sun was fairly active. Meredith, Van Allen, and Gottlieb[10] found no evidence for particles below about 1 BeV/nucleon in 1952 (the Sun was still active), indicating a rather sharp low-energy cutoff. The depression of the cosmic-ray intensity during the years of solar activity is obviously due to the disappearance of the low-energy particles from the spectrum, as is shown directly by the shift of the latitude knee toward the equator; the mean energy of the particles observed at Earth increases with decreasing cosmic-ray intensity.

[8] P. Meyer and J. A. Simpson, *Phys. Rev.* **99,** 1517 (1955).
[9] H. V. Neher, *Phys. Rev.* **103,** 228 (1956); **107,** 588 (1957).
[10] Meredith, Van Allen, and Gottlieb, *Phys. Rev.* **99,** 198 (1955).

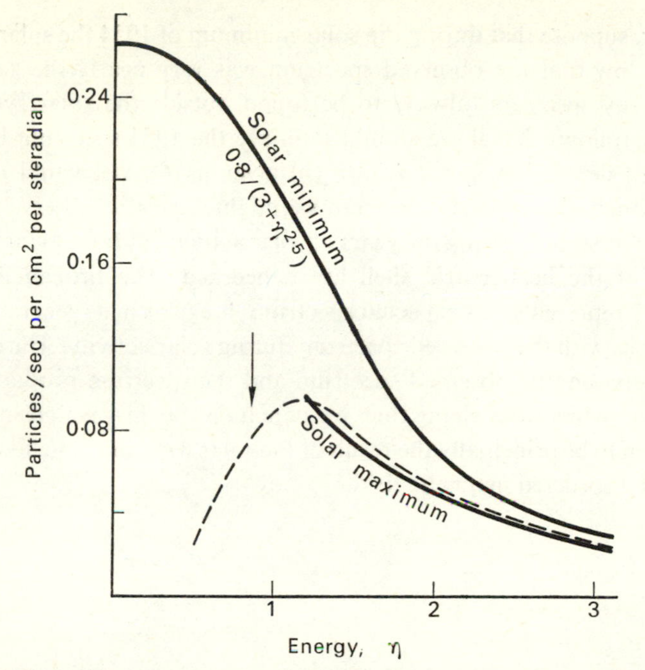

FIG. 2. The solid curves represent the observed primary cosmic-ray differential energy spectrum during the 1954 period of minimum solar activity and during the years of maximum activity; the curve $0{\cdot}8/(3+\eta^{2\cdot5})$ below 1·2 BeV is a qualitative representation of the low-energy observations; the vertical arrow indicates the energy at which the spectrum was observed to drop off rapidly during solar activity. The broken line is the spectrum which results from the solar wind in the heliocentric shell.

We exhibit the observed cosmic-ray differential energy spectrum in Figure 2 during the years of minimum and maximum solar activity. We have used the data of Meyer and Simpson[8] above 1·2 BeV for the solar minimum of 1954 and for the preceding solar maximum. We have included, in a qualitative way, Neher's observation[9] of the high abundance of low-energy particles during the solar minimum of 1954 by supposing that the spectrum is something like $0{\cdot}8/(3+\eta^{2\cdot5})$ particles per sec per cm² per steradian below 1·2 BeV. The vertical arrow at 0·9 BeV represents the energy below which Van Allen *et al.*[10] found a rapidly diminishing particle density during solar maximum.

Now, suppose that during the solar minimum of 1954 the solar wind was so low that the observed spectrum was very nearly the galactic cosmic-ray spectrum (always to be found outside the solar system). Then it follows that if we should multiply the 1954 spectrum by the expected depression factor $j_0(\eta)/j_\infty(\eta)$ given in (6), we would obtain some rough idea of what spectrum we might expect to observe in the inner solar system during the years of solar activity, at least so far as the effects of the heliocentric shell are concerned. The broken line in Figure 2 represents this expected spectrum; it agrees in its general characteristics with the observed spectrum during solar activity. The difference between the observed spectrum and the spectrum produced by the solar wind is so slight that we conclude the 11-year cosmic-ray variation to be principally the result of the solar wind in the heliocentric shell of disordered magnetic field.

Index